Robert Hogg, Henry Graves Bull

The Apple and Pear as Vintage Fruits

Robert Hogg, Henry Graves Bull

The Apple and Pear as Vintage Fruits

ISBN/EAN: 9783337524388

Printed in Europe, USA, Canada, Australia, Japan

Cover: Foto ©berggeist007 / pixelio.de

More available books at **www.hansebooks.com**

THE
APPLE & PEAR
AS
VINTAGE FRUITS.

The technical descriptions of the fruit are for the most part by

ROBERT HOGG, L.L.D., F.L.S.,

Honorary Member of the Woolhope Naturalists' Field Club; Vice-President of the Royal Horticultural Society; Author of 'The Fruit Manual'; *'British Pomology'; 'The Vegetable Kingdom and its Products',* *&c., &c.*

" *Hope on.* "   " *Hope ever.*"

GENERAL EDITOR:

HENRY GRAVES BULL, M.D., &c.,

J.P. for the City and County of Hereford,

MEMBRE HONORAIRE DE LA SOCIÉTÉ CENTRALE D' HORTICULTURE DE LA
SEINE-INFÉRIEURE, FRANCE.

HEREFORD:
PRINTED AND PUBLISHED BY JAKEMAN & CARVER.

1886.

PHILLIPS! POMONA'S BARD! THE SECOND THOU
WHO NOBLY DURST, IN RHYME UNFETTERED VERSE,
WITH BRITISH FREEDOM SING THE BRITISH SONG:
HOW, FROM SILURIAN VATS, HIGH SPARKLING WINES
FOAM IN TRANSPARENT FLOODS; SOME STRONG, TO CHEER
THE WINTRY REVELS OF THE LABOURING HIND;
AND TASTEFUL SOME, TO COOL THE SUMMER HOURS.

THOMPSON, *Autumn.*

PREFACE.

"Quid sit pulchrum, quid turpe, quid utile quid non."
HORACE, Lib. I., Ep. II., 3.

A century has nearly elapsed since any systematic British work has been published on the Apple and Pear, as Vintage Fruits, their varieties, cultivation and management. Marshall first published his work on "Rural Economy" in 1789, and Thomas Andrew Knight's treatise on "The Culture of the Apple and Pear" appeared in 1797. Mr. Knight's last work, the "Pomona Herefordensis," was published in 1811, and in this some thirty varieties of Fruit are so beautifully represented, that it will ever retain its interest. Mr. Knight may be said to have been the first to point out, that the real value of Cider Apples and Perry Pears must be sought in the richness of their juices, as shown by their density or specific gravity; but his studies, in this direction, ended here. The stirring events of the time absorbed all interest, and the profits of agriculture from the growth of cereals, and the production of cattle, threw the Orchards into a state of neglect, from which they have yet to recover. In these days the changes of commerce have again brought Apple culture into consideration, and it has become a matter of importance to attend more carefully to the Orchards, and to bring Science to the aid of individual effort as derived from practical experience.

The first active measures for the improvement of the Cider Orchards were taken on the Continent. *La Société Centrale d'Horticulture de la Seine-Inférieure* appointed a Pomological Committee, presided over by Monsieur C. Lesueur, of Rouen, for the special study of Cider and Perry fruits, which had worked for some years, when in the Spring of 1862, the Society extended its operations by calling to its aid all persons interested in the Orchards. By these efforts, assisted by those of Messieurs de Boutteville, of Rouen, Michelin, of Paris, Thierry, of Caen, and several of the leading pomologists of France, the Government was induced in 1864, to appoint a CONGRÈS POUR L'ÉTUDE DES FRUITS À CIDRE, with its centre of operations at Rouen. This Congress held its meetings successively in the leading Cider districts of France, viz.: at Caen (Calvados) in 1864, where it first took a definite form; at Rennes (Ille-et-Vilaine) in 1865; at Alençon (Orne) 1866; at Beauvais (Oise) 1867; at Saint Lô (Manche) 1868; at Bayeux (Calvados) 1869; and at Yvetot (Seine-Inférieure) in 1871. The results of all these labours were arranged systematically, by Messieurs L. de Boutteville and A. Hauchecorne, and published in 1875, under the title of "Le Cidre." This work is of a highly scientific and comprehensive character. It is thoroughly practical, and it has rendered very great service to the Orchards of Normandy.

The SOCIÉTÉ CENTRALE D'HORTICULTURE DE LA SEINE INFÉRIEURE has continued its labours since this period, with the same energy and perseverance, until at the present time, the Society has nearly four hundred varieties of Cider Apples and Perry Pears, modelled in wax, and carefully coloured to Nature, in the rooms at the Hôtel des Sociétés Savantes, at Rouen. These fruits have all been carefully examined, and their juices analysed. A Catalogue has been drawn up, which gives in a tabular form the name of each variety; its periods of blossoming and of maturity; the flavour of the fruit; the quality and density of the juice, and the amount of Sugar, Alcohol and Tannin it affords; together with a brief notice of the general character and habit of the tree. The Fruits in the Catalogue are divided into Classes according to merit, and for the convenience of distinction a colour is attached to each Class.

The First Class (Cartes Jaunes) consists of "excellent" Apples, and it gives twenty four varieties. The Apples in this Class contain Sugar, Alcohol, Tannin and Perfume in sufficient quantities to yield a rich, long-keeping Cider of excellent perfume and flavour; whilst it retains a sufficient amount of unreduced Sugar, to give sweetness, and enough Tannin, to give strengthening virtues, and at the same time to moderate the action of the Alcohol.

The Second Class (Cartes Blanches) consists of "very good Apples," and it gives fifty one varieties. These Apples yield juices with sufficient Sugar, Tannin, and Perfume to make a rich good-keeping Cider.

The Third Class (Cartes Saummonnées) are "good Apples," and it presents sixty eight varieties. Their juices yield a pleasant Cider very good in flavour, but without much strength, or keeping qualities.

The Fourth Class (Cartes Lilas) consists of "Middling or Bad" Fruits, and it gives two hundred and five varieties. These contain in a very inferior degree, the useful properties of those in the three former Classes.

This Catalogue also gives the results of the enquiries into the virtues of twenty seven varieties of Perry Pears, of which one variety only is put in the First Class; two in the Second; nine in the Third; and Fifteen in the Last Class.

This Catalogue thus affords the most useful and valuable information, as to the real merits of the several varieties of Fruit, in a concise form, and renders great service to the cultivators in the formation of their Orchards.

The Congress also laid down this general rule as the result of their labours, that the minimum density of the juice of Cider and Perry Fruits, should be 1.075, with at least one half per cent. of Tannin.

THE WOOLHOPE NATURALISTS' FIELD CLUB has been engaged during the last nine years in obtaining Orchard information, with a view to improve the varieties of Fruit grown, and to restore the commercial position of their products. The result of all these enquiries is embodied in "The Herefordshire Pomona." This work has been published at very considerable expense; and is very valuable for the carefully coloured illustrations of Plates, containing four hundred and thirty two of the most highly esteemed varieties of Fruit for the Table and the Press. It thus forms an excellent work for reference, but it is far too costly and valuable for general use. The Club has therefore thought it advisable, for the advantage of the Orchards, to publish at once the present cheap Edition of all the information contained in the larger work, with reference to the Apple and Pear, as Vintage Fruits. In this more useful form, the results of the enquiries will be at once available for the improvement of the Orchards; and it is hoped that this work may become the Text Book for practical use with the Nurseryman and the Planter, until a better one is published. It offers no pretensions to the complete and highly scientific character of the French work, "Le Cidre," since the resources of the National Government have not been available here for the long and expensive investigations required, but its enquiries have followed the same paths, and it will at least afford the ground work for future and more perfect results.

NOTE.

The Publication of this Volume is in fulfilment of one of the most earnest wishes of its Editor, who was engaged in bringing the work through the Press, when sudden fatal illness prevented his seeing more than the first portion in print. The work, therefore, lacks the finishing touches, which his experienced hand would have given; but care has been taken that the work should be printed exactly as it was left by him. It is not a mere reprint of the "Herefordshire Pomona"; for much additional knowledge was obtained, and the papers were re-written for this book, that it might be specially valuable to the fruit growers and cider makers of the county.

It was during a visit to Rouen on behalf of the Woolhope Club, on the occasion of the Great Exhibition of Apples and Pears held there in 1884, under the auspices of the *Société Centrale d' Horticulture de la Seine-Inférieure*, to whose work reference is made in the Preface, that the excellence of the Orchards in Normandy was remarked. The care and attention evidently bestowed upon them, and the numbers of young trees planted, were the subject of special notice. It was felt that such results were largely due to the work of the Society mentioned above.

That similar results might be produced in the Orchards of Herefordshire and of England, this work was undertaken by the Editor. It represents the fruit of many years of patient labour and study.

Dr. Bull was greatly indebted to George H. Piper, Esq., F.G.S., for much of the local history of the Orchard fruit, particularly in the neighbourhood of Ledbury; and to J. H. Arkwright, Esq., of

Hampton Court, for kindly issuing circulars with reference to the time of the flowering of Orchard trees, which has proved of value to the work.

Thanks also are due to the Publisher, Mr. Carver, for the great care and zeal with which he has carried the work through the Press.

It will be of interest to know that the Sections of the Fruit were all carefully drawn by Dr. Bull himself

August, 1886.

CONTENTS.

	PAGE
THE ORCHARD AND ITS PRODUCTS—CIDER AND PERRY	1
I.—The Orchard	9
II.—Orchard Trees	20
III.—Fruit Management	37
IV.—Fermentation	46
V.—The Orchard in its Commercial Aspect	73
VI.—Renovation of the Orchards	78
VII.—Orchard Prospects	83
REPORT ON THE CONGRESS AT ROUEN	87
LIST OF THE MOST ESTEEMED VARIETIES OF CIDER APPLES, WITH THEIR SECTIONS	91
LIST OF THE MOST ESTEEMED VARIETIES OF PERRY PEARS, WITH THEIR SECTIONS	177
ADDITIONAL LIST OF LOCAL PERRY PEARS	223
ADDITIONAL LIST OF CIDER APPLES, FROM THE COUNTIES OF HEREFORD, DEVON, SOMERSET, WORCESTER, AND GLOUCESTER	225
GENERAL INDEX	243

THE ORCHARD AND ITS PRODUCTS.

CIDER AND PERRY.

NEC VERO TERRÆ FERRE OMNIA POSSIT.
VIRGIL, *Geor.* II. 109.
"Not every plant in every soil will grow."
Dryden.

"THE FRAGRANT STORES, THE WIDE PROJECTED HEAPS
OF APPLES, WHICH THE LUSTY HANDED YEAR,
INNUMEROUS, O'ER THE BLUSHING ORCHARD SHAKES;
A VARIOUS SPIRIT, FRESH, DELICIOUS, KEEN,
DWELLS IN THEIR GELID PORES; AND ACTIVE, POINTS
THE PIERCING CIDER FOR THE THIRSTY TONGUE."
THOMSON. *Seasons.*

"WOULD'ST THOU THY VATS WITH GENEROUS JUICE SHOULD FROTH?
RESPECT THY ORCHATS; THINK NOT THAT THE TREES
SPONTANEOUS WILL PRODUCE A WHOLESOME DRAUGHT
LET ART CORRECT THY BREED."
PHILIPS' *Cyder.*

The variable and temperate climates of Northern Europe are better suited to the growth of the Apple and the Pear-tree, than to that of the heat loving Vine: and thus in olden times, when communication was difficult, or almost impossible, and when each locality was very much dependent on its own productions, Cider and Perry became the natural drink of the inhabitants. It is not however in every soil and situation that the juice of the Apple and Pear are sufficiently rich to produce fermented liquor of high flavour and quality; and it is curious to observe how limited are the districts to which the experience of centuries has restricted the growth of Cider and Perry Orchards. In England it is only the Western Counties which are noted for their Orchards. The West Midland district, comprising Herefordshire, Worcestershire, and Gloucestershire, with some parts of Monmouthshire; and the South Western district, comprising the Counties of Devonshire, Somersetshire, and part of Dorsetshire. Cornwall also possesses many Orchards; and the fame of Kent is widely spread for its extensive production of dessert and table fruit. In Ireland some fair Cider is made in the Counties of Waterford and Cork, but not to any great extent.

In Normandy, Cider Orchards may be traced back to the 11th Century. They were much more extensively planted between the 13th and 16th Centuries, and now again the destructive disease of the vines is causing the Orchards to be widely extended, so that a considerable quantity of Cider is produced there. Pear Orchards have never been much planted in Normandy, and Perry is but lightly esteemed there. In Germany, on the contrary, Perry is more highly valued than Cider, and is made largely for distillation. Cider has been known in Spain from a very early period. A graphic description of the Cider of Biscay is given by Nasagerus in the Journal of his Embassy from the Republic of Venice to the Emperor Charles V., in the early part the 16th Century; and it now forms the ordinary drink of the inhabitants of the Northern provinces of Spain and Portugal. In Jersey much Cider is made which has a high repute for its strength. In many parts of the United States of America the common drink of the country is Cider; but the manufacture of Perry is chiefly confined to the Eastern States, where it is produced in considerable abundance.

It was not until the end of the 17th Century that the English Orchards began to be much planted. The Civil War with its troubles had passed by: Continental wars prevailed for the most part; and as foreign wines ceased to be introduced, it became an object of national importance—a patriotic duty—to encourage the home production of Cider and Perry in every possible way. Poets and Writers extolled their praise: Esquires and Yeomen vied with each other in their efforts to meet the national want; and the great care and attention resulting from all this enthusiasm culminated in a success so remarkable as to outstrip all former efforts, and as we read the accounts, to make us lament the more, the neglect of later years.

Cider and Perry were then made in large quantities of a more uniform superior quality; and met with a ready and highly remunerative sale. They formed the household family drink, varied on festive occasions with home-made wines, in the excellence of which all good housewives prided themselves. The farm labourers, or hinds, who were at that time usually boarded in the house, had to be content with "Ciderkin," or "Purr," a weaker cider, made by the addition of water to the apple cake, as it was passed again through the mill. This was allowed to the men in almost unlimited quantities during haytime and harvest, and formed a wholesome and harmless drink.

This was the golden age for Orchard culture and for Orchard produce. Cider was never so highly esteemed. Philips, the Cider Poet, calls it:—

> "Nectar! on which always waits
> Laughter and Sport, and care beguiling Wit,
> And Friendship, chief Delight of Human Life.
> What should we wish for more? or why in quest
> Of Foreign Vintage, insincere and mixt,
> Traverse the extremest World; why tempt the Rage
> Of the rough Ocean! when our native Glebes
> Imparts from bountious Womb, annual Recruits
> Of wine delectable, that far surmounts
> *Gallic*, or *Latin* grapes, or those that see
> The setting Sun near Calpe's tow'ring Height.
> Nor let the *Rhodian*, nor the *Lesbian* Vines
> Vaunt their rich must, nor let *Tokay* contend
> For Sov'ranty; *Phanæus* self must bow
> To th' *Ariconian* Vales."—*Cyder*.

This great prosperity of the Orchards was not destined to continue for any lengthened period. Agriculture was soon called upon with greater urgency to meet the want of the more essential articles of food, and it became more profitable to produce corn and cattle; thus the chief attention of the farmer was drawn from his fruit trees and was given to these objects. Orchards are uncertain in their yield; the fruit requires much care and attention, and with all this, a good season is as necessary for superior Cider and Perry, as it is for fine Wines; whereas the grain crops are much more to be depended upon, and the area of their production is practically without limit.

The farmers grew rich, their farms kept increasing in size, and the attention given to their Orchards became less and less, until, at last, they begun to be looked upon sometimes as a nuisance. This neglect, as years went on, became disastrous; failing trees had their places supplied by any worthless varieties at hand; little care was given to the management of the fruit, or to the making of the liquor, beyond the two or three hogsheads required for the household use. Then, year by year, enormous quantities of Cider and Perry of a very indifferent quality were produced, and, as the natural consequence of this deterioration, they could only be sold at prices less and less worthy of consideration. They were, therefore, given the more freely to the labourers on the farm, inducing habits of indolence and intemperance, and, as a matter of course, lessening their wages.

The quantity produced was far too great to be consumed locally, and hence arose the need of the "Cider Merchants," "Cider-men," or "buyers of sale liquors," as they were called at the end of the last century, who bought up everything by wholesale, and almost at their own prices. There can be no question but that, with some honourable exceptions, these "middlemen" have done more damage to the just reputation of Cider and Perry than all other causes put together. In ordinary seasons many thousands of hogsheads passed through their hands, and were submitted to various processes, calculated rather to destroy than to regulate proper fermentation. The liquor was fined, flavoured and fortified to suit, in their estimation, the public taste. It was then sent to London and Bristol, (in those days the two great centres of trade,) the best in bottles to (mis)represent pure wholesome Cider in the home market, whilst the greater part of it found its way, it is said, to the Continent, to return again to this country, in the shape of cheap Hamburgh Ports and Sherries; or, more probably, it was manipulated at home for these purposes. Not a little of this nefarious traffic, it is to be feared, goes on at the present day.

There were other causes also, which tended, from an early period, to lessen the production of Cider and Perry. Taxation was very soon imposed, sometimes on the Orchards, but generally on the produce. This was often very oppressive, and caused many Orchards, not protected by the landlord's agreement or lease, to be uprooted. The obnoxious visits of the Supervisor continued until the commencement of the present century, but have now, happily, ceased for many years.

Foreign Wines soon began again to be introduced during the intervals of war, and their importation has continued to the present time, in ever increasing quantities, with the improved facilities of transport, and the diminution of duty. These cheap wines, aided greatly by malt liquors, have at all times been formidable rivals for public appreciation, and it is a standing proof of the natural excellence of Cider and Perry that they should have been able to hold their own as well as they have done, in spite of so much general deterioration, and in the face of such powerful competition.

The same falling off in the quality of Cider of late years, has been observed in other countries. In France it has been strongly commented on in the official Report of the Congress, appointed by the French Government to consider this subject. This excellent work, "Le Cidre" (pp. 77, 78), says—"The Cider of which the old authors wrote in such glowing terms, is scarcely to be met with now. Such, for example, as the *Ecarlatin*, prepared from the *Pomme Ecarlate* (scarlet apple), which yields an excellent Cider, red as wine, sweet, piquant, and aromatic, as if sugar and cinnamon had been used; or such of the *Muscadel* which recalls the colour, scent, and taste of the *Muscadelle* wine. It is of this cider that the old French soldier-song says—

'Il vant mieux, près beau feu, boire la *Muscadelle*
Qu'allez sur un rampart faire la sentinelle.'

Or lastly, the cider furnished by the apple called *Pomme d'Espice*, which is as superior to ordinary cider as the *Vin d'Orléans* is to *Vin Ordinaire*" It is related, by Julian de Paulmier, that "The late King Francis the Great, in 1532, passing through the district, gave orders that some barrels of it should be carried in his train, and he drank of it himself so long as it lasted." (*Traité du Vin et du Cidre*, published at Caen, in 1589.)

A similar fact of royal appreciation of Cider, is related by Dr. Beale, who wrote in the time of Charles II (1656), and who says: "When the King (of blessed memory) came to Hereford, in his distress, and such of the *Gentry* of *Worcestershire* as were brought there as *Prisoners*; both *King*, and *Nobility*, and *Gentry*, did prefer *Cider* before the best *Wines* those parts afforded."

The same neglect was observed in America, some half century ago, when Thacker called attention to their Orchards. His warning would seem to have been effective, since, of late years, a marked improvement has shown itself, in all kinds of American Apples and Pears, whether for dessert, for culinary purposes, or for the production of Cider and Perry. "American farmers are now beginning," says Mr. Downing, "to recognise the fact, that no farm is complete without a well-selected and well-cultivated Orchard." (*American Fruits for Farm and Garden*, 1871.)

The wonders effected in commerce by the great discoveries of the present century, have completely surpassed the results of all former experience. The power of the steam engine, by land and by sea, enables space to be overcome by rapidity of movement, and lessens expenditure by gain on time, and cheapness of conveyance; and thus wider markets are offered for all articles of trade. Nor have these changes by any means reached their limit. Every year sees some new economy effected, or some fresh article of commerce introduced into new districts to compete with those already in the field. Competition thus becomes world-wide, and according to the inevitable laws of trade, the best and the cheapest must prevail in the end. The benefit to humanity at large is unquestionable, but to individuals and localities the result is often disastrous. Agriculture is now tried severely to contend with these great changes, and the struggle still goes on with increasing severity, in almost all the articles of its production. The result cannot be otherwise than to compel every district, and every locality, to produce those articles for which it is specially adapted, in the best possible form, or, in other words, by the highest cultivation. If free-trade in corn, and the introduction of live and dead meat, restrict the profit of the farmers, happy should they be, who, living in the fruit districts of England, have their Orchards to help them.

Two hundred years ago, it was the necessities of isolation that caused the Orchards to be looked to as a good source of profit; in these times it is a world-wide competition that makes the same demand. Thus it has come to pass, by a curious revolution in the cycle of commerce, that the careful cultivation of English Orchards has again become a necessity, and every effort must be made to improve their condition, and to make them, as they can be made, one of the main sources of the profit of the farm.

The fruit districts in England, in all ordinary seasons, should afford the chief supply to the English markets, but they do not do so. American and Continental Apples and Pears are brought, year by year, in larger quantities, to supply our great centres of population, and they are even now coming from Australia. These importations are always noticed to possess the two leading marketable qualities, "size," and "beauty of colour," and the best are also excellent in flavour and quality. In bad seasons, as in 1879, particularly, American Apples were bought to supply our own apple districts. This competition, will, for the future, always have to be encountered, and it is very satisfactory to know that it may be met successfully, by care and attention to our own Orchards. Of late years, table and kitchen fruit, "pot fruit," as the local name has it, have been much more extensively grown here, and they must still be grown, in increasing quantities, and in improved quality. This particular change, however, will not prove the one great remedy for agricultural prosperity that has been recently claimed for it, for the Cider and Perry fruits must also be grown with increased care, and in improved varieties.

The English Orchards afford a still better resource in their vintage fruits. The products in which they are unrivalled, and for which, therefore, they need not fear competition, are Cider and Perry of superior quality. Here is the speciality that requires the immediate attention of our fruit growers; and it is one that will repay all the care they can bestow upon it. For many years past, the Cider and Perry of first quality has been made by the small holders of land. They have been obliged to look chiefly to their Orchards for their rent and their livelihood; and by unremitting attention to their trees, have received a liberal and just reward. The holders of the larger farms and larger Orchards, must follow their example. It does not answer to produce a drink of inferior quality, when it is possible to produce a better; and it may assuredly be said now, as truly as it ever could have been said, that so long as the quality is superior, however large the quantity may be, a ready market will be found for it, at highly remunerative prices.

The writers of the 17th and 18th centuries, produced many excellent practical works on Orchard culture, and on the manufacture of Cider and Perry. They were, for the most part, the result of personal experience, and vary greatly in their views: indeed they also show signs of local origin. The Orchardist, whose land is variable, and but little of it good, thinks "soil" is the one thing essential; he whose land has been undrained, and whose trees grow unkindly, with rugged moss-covered branches, lays great stress on "drainage"; he whose Orchards are on low ground, and exposed to night fogs, and whose hopes have been cut down again and again by late spring frosts, destroying the fertility of the bloom, dwells fondly on the all importance of a "sunny, airy, upland situation." He whose land is everywhere good, and well adapted for Orcharding, throws all the energy of his recommendations into the absolute need of selecting "the best varieties of fruit" for cultivation; whilst, lastly, he who happily possesses all these advantages, considers that "the management of the fruit, and its proper fermentation," are the requisites supremely essential for the production of Cider and Perry of the highest quality and excellence. All these good people are right from the result of their own experience, but all are wrong in the restriction of their views. The careful personal attention of the cultivator must be given to each and every one of these points, with patience and perseverance, and then it will only remain for favourable seasons to insure a full amount of success.

The present condition of English Orchards is far from satisfactory. They show sadly the result of long-continued neglect. It is the object of the present work to direct attention to them, to give a brief, practical review of the requirements for their proper cultivation and management, and thus to pave the way for further and more complete study.

I. THE ORCHARD.

SOIL.—The Apple and the Pear-tree are very hardy. They will grow and flourish in almost every variety of soil, producing in abundance their most useful fruits. The Apple-tree prefers a Sandstone wherever it is found, as the Pear-tree rejoices in Calcareous soil. It has been universally observed however that the same trees will produce fruit varying much in size and quality on different soils. "Every variety of Apple," says Thomas Andrew Knight, "is more or less affected by the nature of the soil it grows upon. On some soils the fruit attains a large size and is full of juice, on others it is dry and highly flavoured."

When fruit is required for Cider making, the proper quality of the soil on which it is grown is all important. As the poet has well said :—

> "Next let the Planter, with discretion meet
> The Force and Genius of each Soil explore ;
> To what adapted, what it shows averse :
> Without this necessary care, in vain
> He hopes an Apple Vintage, and invokes
> *Pomona's* aid in vain."
>
> <div align="right">PHILIPS " <i>Cyder.</i>"</div>

Happily, however, the rough handed experience of every day life has been able to get on in advance of Science. The practical farmer has not to wait for the chemist to tell him which of his fields are most productive. The dairyman, for example, soon finds out from which of his meadows he gets the best milk, the richest cream, and the most valuable cheese; and his next object is to get the best breed of cattle to graze them, or in other words to find the cows that will best perform their part in dairy produce. So it is with the Orchardists, the liquor in his vats will soon point out to him the particular Orchards which offered him Nature's best laboratory for the production of the finest and strongest Cider; and his efforts should then be directed to get them provided with the best varieties of fruit trees. It is with Orchards moreover, as it is so remarkably the case with Vinyards, that some portions of the ground will produce much finer liquor than the rest, although the soil apparently is the same throughout. The fact is undoubted, but the reason seems inscrutable and beyond the powers of chemistry to define.

The Cider and Perry from the English Orchards are admitted to be superior in quality and strength to those liquors from other countries, and thus our Orchards should show the soil best suited to their production. The evidence from history on this point is not quite satisfactory, for all the authorities of the 17th century agree in recommending light sandy soils, such as are usually termed "Rye Lands."

> "Look where the full-eared Sheaves of Rye
> Grow wavy on the Tilth, that Soil select
> For Apples." PHILIPS "*Cyder.*"

Mr. Thomas Andrew Knight says "the excellence of the Cider formerly made from the *Redstreak, Golden Pippin,* and *Stire* apples in light soils seems to evince that some fruits receive benefit from those qualities in the soil by which others are injured." Marshall gives the instance of the once celebrated *Stire*, which in the limestone lands of the Forest of Dean yielded an incomparably rich and highly flavoured Cider, but when grown in the deep, rich soil of the vale of Gloucester, afforded a liquor only useful for its strength and roughness. The *Hagloe Crab* again, another celebrated apple in its day, required the calcareous rock called "Dunstone" to give full flavour and richness to its liquor. The *Foxwhelp* on the other hand, yields the Cider, so remarkable for its strength, and that peculiar flavour, for which it is so highly esteemed, from deep clay Sandstone loam, and if the trees are grown on light or too sandy a soil, its Cider is then thin and inferior in flavour. The same may be said of several other varieties.

It is a curious fact, and certainly more than a coincidence, that the practical experience of so many generations of men should show that the two counties which have chiefly given its high character to English Cider are Herefordshire and Devonshire; and these two counties are remarkable for the same character of soil, that is for the deep clay loam of that ancient geological formation, the Old Red Sandstone. This experience is fully borne out in our own times, and it may be added that even in these favoured counties, the districts especially noted for this character of soil, are those most remarkable for Cider of the highest flavour and quality. The light soils will not now give a superior Cider, and he who would plant a successful Orchard must choose a deep, stiff, Sandstone loam for his trees, if he has the opportunity of doing so.

The following analysis of Herefordshire soil was made by Mr. G. H. With, F.R.A.S., and F.C.S., in 1877:—

ANALYSIS OF THE CREDENHILL MARL, OR CORNSTONE.

Organic matter and combined water ...	2·261
Silica, and insoluble Silicates ...	56·068
Tricalise Phosphate ...	·391
Lime Carbonate ...	26·098
Magnesia Carbonate ...	2·211
Peroxide of Iron ...	5·170
Alumina	3·600
Chloride of Potassium	1·070
Chloride of Sodium	·427
Peroxide of Manganese, Sulphuric Acid, and loss ...	2·704
	100·000

Credenhill is noted for its Orchards, and their fertility is due in great measure to the supply of Lime from the Marl or Cornstone, which surrounds this hill, as it does so many others in Herefordshire.

The Pear-tree is still more hardy than the Apple-tree. The blossoms resist well the spring frost, and the trees bear abundantly. The celebrated variety, *Taynton Squash*, draws its finest liquor from the heaviest soil; and that popular Pear, *Bare-land Pear*, takes its name from the coldness and poverty of the soil it grows on. Thus it happens that Perry may be produced to great profit and advantage on many a soil that will scarcely give back the labour spent on it for other purposes. Pear trees are very slow and long lived. The The old proverb says—

"He who plants Pears
Plants for his heirs."

and thus the unselfish patriotism which should plant Perry Orchards is not always to be found. However a good "hit" of fruit in an Orchard of Pears has sometimes been *worth the fee simple* of the land the trees grow upon.

SURFACE.—The question of turf, or tillage, as best adapted for Orcharding has been much discussed; and pasturage has been commonly favoured under the idea that the soil beneath the trees was thus kept more cool and moist during the heat of summer. This is not the case; for the crop of pasture, or hay, or green crops of any kind not only require much moisture for their own growth, which they take from the soil, but they also exhale much more moisture during the heat of the day time, compensated for by the dew that falls on them by night; and thus in both ways the trees are robbed in dry weather of the moisture necessary for their healthy and fruitful growth.

The old Orchard writers are therefore right in giving preference to tillage, rather than to pasture land, for the Orchard. Thomas Andrew Knight, and most other Herefordshire authorities, think there is no more suitable place for a young Orchard than a Hopyard; and the most approved method in Kent at the present day, is to cultivate the Orchard as a Hop Garden until such time as the fruit trees are large enough to yield a paying crop The trees profit by the high cultivation, and the protection given to the hops. They grow more freely, bear finer fruit, and yield, it is said, a longer keeping Cider. As the trees grow large the hops must be uprooted and the field laid down to permanent pasture.

In America, roots are almost always grown for the first five years in new Orchards, and the soil deeply ploughed every year at a proper distance from the trees. They consider grain crops as too exhausting and injurious to the soil required for Apples.

The home Orchard attached to most Herefordshire and Devonshire farms must be pasturage of necessity, for the great convenience it affords for the ewes and lambs in the spring, and the ordinary farm animals at all seasons.

DRAINAGE.—A due amount of moisture in the soil is absolutely necessary for the proper growth of the higher forms of vegetation, but it should not be in excess, and above everything, it must not be stagnant. A want of good drainage is fatal to an Orchard. The temperature of water-logged soil is always low. The warm rains of spring run off the surface, without mixing with the cold water left there by winter; and it is very late in the year before the sun can lessen its quantity by evaporation, and impart the all essential warmth to the soil. If water moreover remains long stagnant in contact with any vegetable matter it soon becomes impure by the formation of noxious gasses, and is thus rendered positively injurious to the trees growing there. An Orchard in this condition is a miserable sight; the trees are rugged and stunted in growth, their boughs are weak, covered with lichen, or moss, and can seldom produce much fruit; and yet, for all this, it is a sight by no means uncommon.

A good Orchard must therefore be well drained by art, if not by nature. The excess of water should flow off gradually, so as to leave the soil porous and ready to receive from the atmosphere quickly its own air and warmth. The roots are thus stimulated early in the season and have .time to take up from the soil all the principles necessary for the healthy life and vigorous growth of the trees.

ASPECT, CLIMATE, AND SITE.—The Aspect and Site of the Orchard involve its Climate, and there is no subject on which the writers of the 17th and 18th centuries differ more, for though all agree in preferring the South, they embrace also nearly every other point of the compass. The "Complete Planter and Cyderist" (1690) recommends a South, South East, or South West Aspect protected from the North, North East and North West winds by buildings, woods, or high ground. Dr. Beale in his "Tract on Herefordshire Orchards" (1656) preferred a South Aspect inclining rather to the rising than to the setting sun. Mortimer in his "Husbandry" recommends any site from East to West. Mr. Thomas Andrew Knight also thought any Site from East by South to West favourable for orcharding.

The general belief is that the Southern Aspect with an inclination to the East is best adapted for the Orchard, thus following the popular idea of the health giving powers of the morning sun; in other words that this aspect gives a better supply of light and heat, and therefore affords a better promise of healthy vegetation and fruitful crops. This belief holds good for Herefordshire, where the West winds are apt to prevail with much violence, but apart from such special circumstances, any Southern aspect tending Westward is the proper one for an Orchard. It is well known that if plants are exposed to the direct influence of the rising sun at the time they are frozen they will suffer, and in some cases altogether perish; but if the same plants are shaded until gradually thawed by the increasing temperature of the air, they recover from the effects of the low temperature without injury. Hence it is that an Orchard exposed to the direct influence of the morning sun is almost sure to suffer from a spring frost when the trees are in blossom, or when the fruit is setting; whereas with a Western Aspect which does not receive the direct rays of the sun until the increased temperature of the air has dispelled the frost, the blossoms escape and the fruit crop is saved. One side of an Orchard, or one side of a tree is frequently found bearing fruit abundantly whilst the other side is almost bare, and this generally arises from the same cause. If frozen blossoms could be shaded till the sun had diffused its warming influence through the air, and thus had gradually dispelled the frost before its direct rays reached them, the blossom would be saved.

It is sometimes found advantageous to have plantations in different aspects, so as to secure crops in variable seasons. Marshall had an Orchard with a North West aspect which fully fruited in 1783, when the Cider fruit failed in every other aspect. The same fact was experienced in 1879 by Mr. Hill, of Eggleton, and some other growers.

Orchards are often planted too low in the valleys, for though they may get there a more rich alluvial soil and better protection from wind, they have to encounter the cold damp fogs of night, which are often destructive to the blossoms in spring, and are apt moreover to check the free growth of the young fruit after it has set. The best situation, when the soil is good, is one that is raised well above the level of the night fogs from the low ground.

Worlidge has these quaint and consolatory remarks on the best position for the Orchard: "for the distinguishing thereof there are many rules, but he that is seated and fixed in any place, and cannot conveniently change his habitation, must be content with his own, and if any defect or disadvantage be in it, it may be that he hath some advantages that others want."

Wherever the Orchard may find itself, it is desirable to give it the protection of buildings, high quick hedges, woods, or higher grounds to keep off the dangerous spring frosts and blight, and afford as much shelter as may be from strong winds; for then the blossom is often saved from destruction, and the crop of fruit when full grown kept secure.

MANURING.—Apple and Pear trees, whether in arable land or pasture, are very insufficiently manured. The trees often become weak and exhausted from the heavy loads of fruit they bear, and yet their ungrateful owners forget to feed them. This neglect, no doubt, often gives the explanation why so many trees only bear fruit on alternate years. On arable land they take a share of the manure supplied for the green crops grown thus; but on pasture land they have most commonly only to share with the grass the manure from the animals that graze beneath them and enjoy their shade. A careful farmer in the neighbourhood of a town may sometimes scatter a few ashes over the Orchard to help the grass, but it very seldom occurs to him to think that the trees would be equally grateful for some better nourishment.

The kind of manure best suited for the Orchard may be learnt from the consideration of the solid constituents of the tree itself and its fruit, since this analysis must show the inorganic ingredients they demand from the soil. Professor Emil Wolff, of the Royal Academy of Agriculture, Hohenheim, Wirtemberg, has made the most careful investigation into the ingredients of the ashes of plants, and he has published the following results:

ANALYSIS OF THE ASH OF APPLE TREE WOOD.

100 Parts by Weight, gave of

Potash	12·0
Soda	1·6
Magnesia	5·7
Lime	71·0
Phosphoric Acid	4·6
Sulphuric Acid	2·9
Silica	1·8
Chlorine	0·2
	99·8
Loss	2
	100

ANALYSIS OF THE ASH OF THE APPLE ITSELF (whole fruit.)

100 Parts by Weight, gave of

Potash	35·7
Soda	26·1
Magnesia	8·8
Lime	4·1
Phosphoric Acid	13·6
Sulphuric Acid	6·1
Silica	4·3
	98·7
Undetermined Matter, and loss	1·3
	100

Professor Wolff has also given the following results of his examination of the fruit of the Pear:

ANALYSIS OF THE ASH OF THE PEAR (whole fruit.)

Potash	54·7
Soda	8·5
Magnesia	5·2
Lime	8
Phosphoric Acid	15·3
Sulphuric Acid	5·7
Silica	1·5
	98·9
Undetermined Matter, and loss	1·1
	100

The amount of Phosphoric Acid contained in Apples and Pears is shown by these analysis to be so considerable, that these fruits have been considered as specially adapted to sedentary men, who work with their brains rather than with their muscles; for Phosphorus is thought to be the best brain food. However this may be, it has thus been demonstrated that the essential inorganic ingredients for the healthy growth of the trees and their fruits are: Potash, Lime, Soda, Phosphoric and Sulphuric Acids, and that these must all be contained in good Orchard soil; but the mode in which they act and re-act on each other, so as to present themselves in a soluble form that can be taken up by the rootlets—to be again modified by the action of the air in the leaf structure—is not clearly known. Science tells us these principles must be furnished to the plants by the soil, and experience proves the necessity of supplying the loss to the soil, and the great advantage of doing so, by the increased health and fruitfulness of the trees.

The best means for replenishing the soil with these materials is not difficult to point out, but they are not always readily to be obtained on the spot. The ordinary farm-yard manure is deficient in Potash and Phosphates. It is too stimulating, and therefore more likely to cause the production of weak succulent wood than of hard fruit-bearing spurs; and this manure is all wanted, moreover, for the green crops on the farm, and for these it is eminently suitable.

There should be a special corner near every farm homestead especially assigned to Orchard Manure. Its foundation might well be road scrapings, and parings, with ditch and pond cleanings mixed freely with lime, and to this should be added the refuse cake from the cider mill (or "must" as it is wrongly called in Herefordshire). This material is useless for any other purpose, and now only burnt or wasted, should always be returned to the Orchards. It is not great in quantity, but it would always serve to indicate the Orchard Manure heap.

The following materials will be found admirably adapted for Orchard fertilization, whether to encourage the vigorous growth of young trees, or to restore the weak and exhausted state of those which have borne large crops of fruit :

Bone Dust	1 part
Pure dissolved Bone	1 part
Kainit	2 parts
Charcoal dust, or fine Coal Ashes ...	20 parts

They should be well mixed and lightly forked into the surface of the soil around the trees.

Mr. With has supplied the following formula to the Hereford Society for Aiding the Industrious, and it has been published under the name of "With's Universal Manure":

Take of	Cwt.	Qrs.	Cost (about) £	s.	d.
Finely sifted Dry Earth	15	0	0	3	0
Finely sifted Coal Ashes ...	10	0	0	3	0
Kainit, finely powdered ...	0	3	0	3	6
Nitrate of Soda finely powdered	0	3	0	12	0
Best Peruvian Guano	2	0	1	8	0
Best Bone Meal ...	1	2	0	12	0
Pure Dissolved Bone ...	3	1	1	12	6
Super-phosphate of Lime ...	1	0	0	6	0
Coprolite, or Phosphorite Powder ...	1	0	0	4	6
	35	1	£5	4	6

The ingredients must be of the best quality and thoroughly mixed together. The compound should be passed through a quarter inch screen. The cost per ton at present prices, including labour, will be about £3 5s.; and something less than half a ton per acre, every third or fourth year, would suffice, since its effects will be found very durable.

PLANTING.—The young trees selected to furnish the Orchard should be stout and well grown, and not less than 8 or 10 years old. They should be planted at equal distance from each other at spaces varying from 15 to 40 feet apart, according to the habit of growth of the variety, or to the further use it is proposed to make of the ground. Mr. T. A. Knight was in favour of close planting whether in arable, or pasture land. Those planters who wish to have the largest return at the earliest period, should plant the trees at 15 feet apart in the rows, cutting away every other tree as soon as they approach each other, taking care to keep the rows 30 feet apart from each other. Dr. Beale advises that the crab stocks "be settled 30 feet apart, and after three years let the artist be sent for to graft them with the best fruit." Mortimer would have "all trees and rows at 40 feet apart and pruned to grow like a fan." The trees certainly should stand so clearly apart from each other as to allow of their full growth, since a large tree will supply not only more, but better fruit than a small one. They should be planted carefully in lines for the convenience of cultivation, and their roots should be kept as near the surface as may be; that is, they should not be planted too deeply in the ground. The soil beneath should be double dug, and if some roughly broken bones could be put in at the same time, say a peck to a tree, they would form an enduring support to the young trees.

Trees of a similar variety, or of a similar habit of growth, and which ripen their fruit at the same period should be planted together; for thus there will be a greater certainty of uniform space for light and air; the general appearance of the Orchard will be improved; and much time and labour will be saved in gathering the fruit in Autumn. It is better also to have a mixture of early and late blooming varieties in the same orchard, so that if a part of the crop is cut off by any adverse circumstances, such as frosts, storms, or blight, there may be still a chance of saving some portion of it.

When the trees are planted they should be well staked, and if in pasture land, they should be strongly protected from cattle or sheep; and lastly, the Orchard itself should be well fenced in, for it is but too often an inclosure only in name, and its fences badly kept and much trespassed on.

II. ORCHARD TREES.

> "Let sage Experience teach thee all the Arts
> Of Grafting and In-eyeing; when to top
> The flowing Branches; what Trees answer best,
> From Root or Kernel."   PHILIPS " *Cyder.*"

It is the common result of experience in all countries, and on every soil, that the quality of the Cider and Perry manufactured depends very greatly on the particular varieties of Apples and Pears cultivated. It was Mr. Thomas Andrew Knight's opinion that "Herefordshire is not so much indebted for celebrity as a Cider county to her soil, as to her valuable varieties of fruit." So too does the French Commission in its admirable Report, "*Le Cidre*," lament, again and again, the absence in these days of that intelligent industry in the selection of the best varieties of fruit for cultivation, which so distinguished the planters of last century. There is much force in these observations, though they do but present a onesided view of the true cause of the decadence in the quality of Cider and Perry. The present state of our Orchards is most unsatisfactory in this respect, since they contain so large a proportion of varieties which are without name, wanting in character, and it must also be added, failing in merit.

SEEDLINGS.

> "An innate Orchat every apple boasts."
>   PHILIPS " *Cyder.*'

Every Orchard farm, properly cared for, has a nursery for

young trees in some out of the way corner, well protected and well looked after. Young Crab stocks are reared from the kernels left uncrushed in the cake from Crab Apples, after the verjuice has been made. The young plants spring up, and after a few careful transplantings, in five or six years become strong enough to graft with varieties of fruit, whose merits are established.

The most approved method is to separate the pips from the cake by washing, so as to obtain clean seed. Mix this with moist sand, or light mould, and set it aside until February. Then sow thinly in drills, an inch deep, on a firm well manured soil, made as for an onion bed. A few vegetate immediately, but most of the kernels will remain a year in the ground before the young plants appear. The seedlings will grow unequally, but at the end of the second year will generally be ready to transplant into rows a foot apart, and three or four inches from each other. Here they must remain for two years, when the best plants will be strong enough to plant out in the nursery in "quarters," as it is termed, that is on ground well trenched, two spades deep, and heavily manured. The rows must now be two feet and a half apart, and the young trees one foot from each other, when they will be ready for budding the following August. Seedlings should always be transplanted early in Autumn, as soon as the leaf falls, and never later than the beginning of November.

Young seedlings are very commonly grown from the Apple kernels in the cake thrown aside from the cider mill. These young Apple seedlings spring up often unsown. They are planted out, and beyond question often escape grafting altogether. They are left where they grow, and if they are found to bear a good looking "eyeable" fruit they get planted out to supply the vacancies that are so constantly occurring in the Orchard from one cause or other. It is owing to this careless practice that worthless varieties are now found to prevail so extensively.

Those who plant Apple pips or kernels with the view of producing new varieties of fruit will find the process tedious.

Jam quæ seminibus jactis se sustulit arbos
Tarda venit, seris factura nepotibus umbram.

VIRGIL, *Geor.* II. 578.

But slowly comes the tree which thou hast sown
A canopy for grandsons of thine own.

BLAKEMORE, *trans.*

Mr. Thomas Andrew Knight found from his experience that Apple-tree seedlings took from five to twelve years to come into bearing; whilst Pear-tree seedlings do not bear fruit until they are from twelve to eighteen years old. Seedling fruit trees moreover are for the most part worthless, and they should never therefore be planted out in the Orchards until their value has been tested very carefully. The direct and only satisfactory manner of doing this is to examine the juice of the ripe fruit by the Saccharometer, which will show its richness by its density. The result is so rarely favourable that much patient perseverance is required. A special exhibition of Seedling fruit trees was held at Yvetot in Normandy, when 172 selected varieties were sent for examination. Nine only of these furnished a rich juice of high density. Again, Monsieur Legrand of Yvetot, out of 65 carefully grown Seedlings, obtained only one single variety worth cultivating. Mr. Thomas Andrew Knight met with the same result, for amongst the many thousands of Seedlings he grew, few indeed proved to be of any value.

The advantages of Seedling trees are very great. They are more robust and hardy, and consequently they bear more freely, and difficult as it may be to obtain good ones, they must still be grown. It is the right way to obtain new varieties of excellence. The attempt is always interesting, and a philosopher has said that "he who provides a new fruit renders a greater service to mankind than he who wins a great battle." It does require great patience and perseverance, and unselfish fortitude too, for it is not every one who could bear with trustful equanimity to be told that the Seedlings he has grown himself, and watched and petted for years, are worthless as varieties, and good only as stocks for grafting.

BUDDING AND GRAFTING.

> Of every suit
> Graffe dainty fruit.
> Graffe good fruit all
> Or graffe not at all.
>
> TUSSER (1620).

Budding is much more practised in these days than formerly. It presents greater economy in material, in labour, and above all in time. Young Seedlings may be budded about the 3rd or 4th year, and if in the following spring the buds should have failed, they can be grafted, and the chance of blanks on the bed be greatly diminished. Budding and Grafting should both be practised in the nursery, where the growth of the Scions may be well protected and regularly superintended. The young trees should not be allowed to take their places in the Orchard until they are strong in the stem, with a good out-line of head, and this cannot be looked for before the 10th or 12th year of the age of the stock.

A custom has arisen in the Orchards of late years, which is often practised with good effect. It is to regraft trees which show a diminution of fruitfulness, or are altogether unproductive, although they may have attained a considerable age. The Scions should be of some strong variety which succeeds well in the locality, and they should be grafted as near to the end of the branches as possible. They will want careful protection from the wind, but if this is given they come quickly into bearing.

OLD VARIETIES OF ORCHARD VINTAGE FRUITS.—The names of those varieties of Cider and Perry Fruits which were held in the highest esteem during the last two centuries have been handed down to us in prose and verse. The following great Orchard authorities, DR. BEALE, writing in 1657; WORLIDGE, 1675; EVELYN, 1706; PHILLIPS, 1706; HUGH STAFFORD, of Pynes, 1753; MARSHALL, 1789; KNIGHT, 1808, and other writers, give the following apples their highest praise. Amongst the earliest in general repute in

Herefordshire was the *Gennet Moyle*, as renowned too for its cooking properties, as for its Cider. This was soon eclipsed by the *Redstreak*, with its varieties, *Summer, Winter, Yellow, Moregreen*, and *Red*. Evelyn and Philips wrote the *Redstreak* into higher favour than has perhaps been awarded to any other apple:

> "Let every tree in every garden own
> The *Redstreak* as supreme whose pulpous fruit,
> With gold irradiate, and vermillion, shines."
>
> <div align="right">PHILIPS, <i>Cyder</i>.</div>

The *Bromsberrow Crab* from Worcestershire, and the *Westbury Crab*, a Hampshire apple; The *Whitesour, Blackamore, Mydiate, Dufflin, Bitterscale, Great White Crab, Deans Apple*, and *Royal Wilding* from Devonshire; the *Arier, Otley, Olive* and *Coleing* from Shropshire; the *Meriot Ysnot, Lings*, and *Peleasantine* from Somersetshire; the *Heming, Hagloe Crab, Bromley*, and *Forest Styre* from Gloucestershire; and the renowned *Foxwhelp*, first mentioned by Evelyn as coming from the Forest of Dean, and which has since surpassed all others in repute. They also name with much favour, *Woodcock; Friar; Pawson; Oaken Pine: Stocking Apple; White, Red*, and *Green Musts; Summer* and *Winter Fillets* or *Violets; Cowarne Red; Underleaf; Garter Apple; Best Bache; Bennet Apple; Elliot; Coccagee; Dymock Red; Skyrme's Kernel; Woodsell; Joeby Crab*, and *Steads Kernel*. Most of these old writers also mention the *Pearmains* and *Pippins* in great variety, of which the most celebrated, even in those days, was the *Golden Pippin*, as well for the long life of the tree, as for the long keeping of its Cider; *John Apple*, or *Deux Ans; Golden Harvey; Nonsuch; Mangold*, or *Onion Apple; Summer*, and *Winter Queening*, &c., with "all, both *Russettings* and *Greenings*, which have a relish of agreeable Piquancy and Tartness."

The varieties of Vintage Pears named by these great Orchard writers, are the *Barland; Horse Pears, Red and White;* divers *Choke Pears*, whereof the red-coloured yielded the strongest liquors; *Taynton Squash;* The *Red* and *Green Squash; John Pear; Money Pear; Lullam Pear;* and some others with local names.

The researches of the Woolhope Club during the last nine years has fully proved that many of these varieties, formerly so highly esteemed, were either altogether lost, or had almost disappeared from the Orchards. The neglect to cultivate these valuable varieties is, doubtless, very much to be attributed to the prevailing belief, that, "Sorts die out of necessity," or as Mr. THOMAS ANDREW KNIGHT expressed it, "There was no renewal of vitality by the process of grafting, but that the scion carried with it the debility of the tree from which it was taken," or in other words that grafted trees will not live longer than the original tree from which the grafts were taken. This opinion, which still prevails very much in the Orchards, is not however correct. It is found to be wrong by careful observation; it is opposed to the general laws of vegetable physiology; and indeed it is now generally admitted by modern Horticultural Science, that any variety of apple may be indefinitely prolonged with proper care and skill.

The notion that a graft can live no longer than the tree from which it is taken seems to rest upon the assumption, that the new wood which grows from the graft is not a new tree, but only a detached part of the parent. This is evidently a mistake. A branch produced from a graft is as distinctly a new and separate individual, as a branch produced by a cutting. In both cases the bud is the source of the new growth; and physiologically speaking a seed itself differs little from a bud, except in being more carefully protected, and in being spontaneously detached. The embryo in a seed, the bud inserted in budding, the buds in a graft or in a cutting, differ only in their position; and each as it developes, becomes a new individual, and not a mere dependent portion of the parent. The embryo of the seed does undoubtedly give that mysterious rejuvenescence of life, which the bud does not, but in each case the new plant has an independent existence, a distinct and separate life. It inherits more or less of its character from the parent tree, but is nevertheless capable of being largely influenced by the circumstances of its own position.

The Woolhope Club resolved to put the question once again to the test of practical experience. Mr. RICHARD CARINGTON, of St. John's Nursery, Worcester, at the request of the Pomona Committee, kindly undertook to conduct the experiment with three good old varieties of fruit which were almost gone. The result is shown by the following "SPECIAL REPORT" to the members of the Club, which was issued in June, 1883 :—

"The Pomona Committee have the great satisfaction to inform the Members, that the experiments they have caused to be carried on during the last four years, for the restoration of those valuable orchard fruits, the *Foxwhelp* and *Skyrme's Kernel* Apples, and the *Taynton Squash* Pear, have completely succeeded. They have now upwards of 800 young trees in vigorous health, viz :—

	Foxwhelp.	Skyrme's Kernel.	Taynton Squash.
One year maidens, about 3 ft. high	500	100	30
Two year's old trees, 4 to 5 ft. high	80	30	18
Standard *Foxwhelp* trees, 5 to 6 ft. high	100		

These young trees have been distributed through the county, and so far as can be judged at present are doing well."

The difficulty of procuring true grafts of the old noted varieties is often very great; for example, it was not until 1883 that they were able to obtain grafts of those valuable fruits, *Forest Styre* and *Hagloe Crab*. They were obtained at last through the kindness of that excellent practical pomologist, Mr. WILLIAM VINER ELLIS, of Minsterworth, near Gloucester, who sent both fruit and grafts, and these excellent varieties are now being propagated by Messrs. CRANSTON & Co., of King's Acre Nurseries, Hereford.

MODERN VARIETIES OF ORCHARD VINTAGE FRUITS.—The new varieties of Cider Apples and Perry Pears introduced into our Orchards during the present century are very numerous. Several of the old varieties already mentioned still remain with us and retain this renown. The *Foxwhelp* which has been the favourite apple for nearly two hundred years still lives and is propagated, and this is also the case with several others of the old varieties, such as *Dymock Red; Cowarne Red; Bromley; Styre Wilding; Skyrme's Kernel; Forest Styre; Woodsell; Joeby Crab; Hagloe Crab; Elliot; Garter Apple*, with others of local repute.

MODERN VARIETIES OF APPLES.

Many valuable additions have also been made of late years, although the history of their appearance and their distribution in the Orchards cannot be clearly ascertained. Devonshire adds *Kingston Black, Golden Bittersweet, Netherton Late Blower, Alford, Sweet Buckland, Devonshire Redstreak* and many others. Somersetshire has added to her list, *Horner,* or *Hangdown; Northwood Bittersweet; Soger; Red Cluster; Tremlett's Bitter; Lopen Neverblight; Jersey, Chisel* and *Flenier; Farmer Hearland; Langworthy's Sour* and *Sweet Natural; Morgan's Sweet;* and various local varieties. Gloucestershire rejoices in *New Bromley; Red Royal; Ansell; Rusty Coat; Maggie; Morris' Pippin; Grittleton, Red* and *Yellow;* and others. Worcestershire adds *Yellow Styre; Cider Brandy Apple; Golden Worcester; Ramping Taurus; Red Splash;* and many local varieties. Herefordshire also presents many novelties of value, such as *Eggleton Styre; Royal Wilding; South Queening; Cider Ladies Finger; Green Wilding; Black Wilding; Pym Square; Munn's Red; Yellow, Spreading,* and *Upright, Redstreaks; Wilding Bittersweet; Bran Rose; Red Styre; Cook's Kernel; Reynolds' Crab; Knotted Kernel; Carrion; Golden Moyle; Red Bud; Black Bud; Tanner's Red; Pin Apple;* and many others, which for the most part have but little merit.

Several of the most valuable apples introduced into our Orchards during the present century, have not been alluded to in this list, they are the so-called "Norman" apples. A great doubt has been felt for some years, as to whether these "Norman" apples were really varieties from Normandy, and every effort has been made to ascertain their history and origin.

Marshall in his book on "Rural Economy" (1789), in the chapter on "Herefordshire Orchards," first notices the fact, of the name "Norman" have been given to a *Wilding* growing in a hedgerow near Ledbury. He very properly points out the error; but from that time, notwithstanding, the custom seems to have prevailed more and more, until of late years, all seedlings, or other unknown fruits, especially if they are "Bittersweets," have had the name "Norman" attached to them. The absurdity is very glaring, when the varieties are named after Englishmen, as *Barnett's Norman, Hawkins Norman, Phillips Norman, &c.;* or from English villages, as *Cummy Norman, Didley Norman, Marden Norman, &c.;* and equally self evident is the anomaly, when such names as "*American Duke,*" *Duke of Normans, Pride of Normans, &c.*, are given to them.

There are nearly twenty of these so-called Norman apples in our Orchards, and several of them have become well-known through the county, and are highly esteemed. It was resolved to compare them with the real apples of Normandy. In the year 1883, through the great kindness of Monsieur FÉLIX DENNIS, a merchant at Hâvre, a very fine collection of cider apples was obtained direct from Normandy, and sent to Hereford. Eighty-five of the best Norman varieties were exhibited in the Woolhope Room, at the Free Library; but not a single one of them was similar to any of the Herefordshire fruits. In order to complete the experiment, it was necessary to take the first opportunity of placing these so-called "Norman" apples of this county upon the tables in Normandy.

Last year, 1884, a grand Congress of the Pomological Societies of France was announced to be held at Rouen. THE SOCIÉTÉ POMOLOGIQUE DE FRANCE, in conjunction with the ASSOCIATION POMOLOGIQUE DE L'OUEST decided to hold its Session at Rouen, from October 2nd to the 12th; with the co-operation of the SOCIÉTÉ CENTRALE D'HORTICULTURE DE LA SEINE-INFÉRIEURE and other kindred Societies from the Départments of LA MANCHE, ILLE ET VILAINE, &c. Exhibitions of Table Fruits and Vintage Fruits were also held, including Cider, and all other Orchard products and Orchard Machinery. An invitation was sent to the Woolhope Club to attend the Congress, and a Committee consisting of Dr. HOGG, of London; Mr. GEO. H. PIPER, of Ledbury; and Dr. BULL, of Hereford, was appointed to represent the Club at Rouen and to compare the Fruits of Herefordshire with those of Normandy.

The ability and energy with which these gentlemen carried out their duties may be almost said to have added an international feature to the Congress. The Report of the Committee is attached to this paper. Its success was most gratifying, and, as will be seen, a Gold Medal was awarded to the Herefordshire Table Fruit; a Bronze Medal to the Vintage Fruit; a Silver Gilt Medal to the Cider from mixed fruit, and a Silver Medal to Cider made from a single variety of Apples; and a large Silver Medal was also given to a bunch of Black Alicante Grapes from Eastnor Castle. To the parts already published of the present work, the high reward of a "Diplôme d'Honneur" was given from each of the Societies under whose auspices the Exhibition for Table, and Orchard fruits, were held. The very high personal compliment of a Gold Medal, was also given to Dr. HOGG for the great services he has rendered to Pomology.

In the comparison of the Orchard Fruits of the two countries the labours of the Committee were also very effective and practical. They have proved as far as possible, that the so-called Norman apples of Herefordshire are not really Norman fruits; and it may be added, that the result of a long series of enquiries renders it almost uncertain, that they are mere local seedlings. The conclusion therefore is, that wherever the name "Norman" has hitherto been attached to a descriptive prefix, it should at once be changed into "Hereford;" and where it is attached to the name of an English person or an English place, it should be changed to "Kernel," or "Seedling." The following varieties, which were exhibited at Rouen, will therefore lose their Norman appellation, and assume the following names:—

BLACK HEREFORD.	RED HEREFORD.
BROADLEAVED HEREFORD.	SPREADING HEREFORD.
BROWN HEREFORD.	SHORTJOINTED HEREFORD.
CHERRY HEREFORD.	SQUARE HEREFORD.
GREEN HEREFORD.	STRAWBERRY HEREFORD.
HANDSOME HEREFORD.	SWEET HEREFORD.
HEREFORD BITTERSWEET.	UPRIGHT HEREFORD.
HEREFORD REDSTREAK.	YELLOW HEREFORD.

The right name of the apple hitherto called *White Norman*, is WHITE BACH, which it must retain; *Phillip's Norman* should be PHILLIP'S KERNEL; *Marden Norman*, MARDEN SEEDLING, and so on for all varieties bearing the names of English persons, or English places.

The great care with which the Committee carried out their next very important duty, that of selecting some of the best Norman Apples to introduce into Herefordshire, is shewn by the Report. The apples they have selected are ROUGE BRUYÈRE, BRAMTOT, MÉDAILLE D'OR, BEDAN-DES-PARTS, MICHELIN, ARGILE GRISE, DE BOUTTEVILLE, and FRÉQUIN AUDIÈVRE. Sections of these several apples with the history, description, and analysis of each of them, will be found in the body of this work.

The PERRY PEARS now most in favour in our Orchards are the varieties, *Taynton Squash; Thorn Pear; Barland; Yellow* and *Black Huffcap; Moorcroft* and *White Moorcroft;* the *Longlands, Old, Winnal's,* and *White Longland; Chaseley Green; Aylton Red; Butt Pear; Red Pear; Thurston Red; Rock Pear* (a late Worcestershire pear of the highest value), *Dymock Red; Turner's Barn*, and there are several other varieties of local repute.

There can be no question, but that there is a very large percentage of Vintage Fruit trees in the Orchards at the present time, which should be "grubbed up" as the country phrase hath it, if they are past the age of regrafting at the ends of their branches with better varieties. They are useless for making good cider themselves, and they serve now but to spoil that which is made from other and better apples. The fruit is unsaleable for these reasons, and it would be economy in every sense to turn the trees into faggot wood.

ORCHARD PRUNING.—The necessary pruning in the Orchard is very apt either to be neglected altogether, or to be carried out in excess. In the one case the boughs grow matted together, and bear their fruit small in size, and deficient in quality; or in the other, whole boughs are mercilessly lopped off close to the trunk, leaving those great round scars, commonly called "Owls faces," to offend the eye of every good Orchardist; since they show how deeply the trees have been injured. It would sometimes seem as if the want of faggots suggested "a turn at pruning," when the poor trees are mercilessly attacked, at the cost of their strength and vigour.

Apple and Pear trees when full grown require very little pruning. "The compleat Planter and Cyderist" says well "while your tree is young, bring it into a handsome shape and order, and when it comes to bear fruit forbear pruning, unless in case of broken, or such broughs as grow cross, or gall and fret others." Mortimer gives similar advice and adds "thin most of the outmost branches, or where they are thickest." Thomas Andrew Knight also lays great stress on judicious pruning, for he did not fail to observe the injury done in the Orchards from the wholesale lopping off of great branches. The scar does not get covered, it decays, and the tree becomes hollow and is broken off by the wind, or split down the middle; and the term of its natural life is materially shortened; and yet it is not difficult to remove large branches without injury, if it is properly done.

The late Mr. Chandos Wren Hoskins in a paper on "Pruning" in the Woolhope Club's Transactions for 1867, has so well explained the true principles on which Pruning should be done, that a short abstract of his paper will be useful.

"The trunk of a tree is fed by its branches, just as a river is fed by its tributaries. It is not nourished by the sap taken up by the roots from the soil, until it has been acted upon by the atmosphere in the leaves; and thus its growth is downward from the foliage, and not upwards from the roots. Every branch of a tree has smaller branches of its own, and is in fact to them a tree. Now, supposing a branch to be condemned, instead of proceeding by capital punishment (which admit of no repentance *except to the inflictor*), the humane process is this. Select a branchlet which happens to grow in the most favourable direction, and at the point where it springs, cut off the main branch obliquely in the direction of the growing branchlet, undercutting at first to prevent spaltering, and prune the wound as much as possible into symmetry with the direction of the new leader. In another year or two serve the new leader in the same way, and the process may be repeated if requisite. The result is this. The growth of the original condemned branch is entirely stopped, without its being itself killed, and as the trunk of the tree increases, its size gets less in proportion, and may generally in a few years be moved entirely without injury, or eye sore, close to the stem, that is to say, when the proportionate size of the scar to the stem is such, that it will heal perfectly in two or three summers."

Trees grow in very different forms, some are upright, some spreading, some straggling in growth, and others altogether irregular. The careful pruner will take the peculiarities of every variety into consideration and leave in each as much bearing wood as possible, always remembering the great physiological truth, that in a healthy tree the extent of root surface must be balanced by the extent of foliage, to produce a well grown fruitful tree. Mr. T. A. Knight, deplored the system of pruning in his day, which consisted in eliminating every branch in the middle of the tree until at length "small tufts of branches were left at the extremities of long and large boughs." This is not altogether the fault of the pruner, for in the growth of spreading mop-headed trees, the middle of the tree is thrown completely in shade, and the smaller boughs if not removed, could never bear healthy fruit. It is more commonly the result of having them planted too closely together in the Orchard. Cutting off main branches should only be required in young trees, and when this is rightly done, no leading branch should afterwards be touched, and the trees should be left to live out the natural term of their lives and fruitfulness.

FRUIT TREE ENEMIES.

A volume might be written on the many enemies that attack Apple and Pear trees in health and disease, and without much avail since few of them admit of the ready application of any remedy. A brief notice must yet be given of those which most commonly and persistently affect them, such as : Mistletoe, Canker, Insects, Fungus, and other vegetable parasites.

MISTLETOE *(Viscum Album)*.—The health and vigour of the trees in an Orchard will generally denote the amount of attention given to them by the owner; but neither care nor attention can keep off mistletoe in a Herefordshire Orchard—Thrushes and other birds eat the Mistletoe berries. The seeds they contain, pass through their bodies, and are thus sown on the branches of the trees they frequent. The young seedlings send their roots into the tissues of the tree, and live at its expense for the future. There is a common impression amongst Orchardists that the Mistletoe renders the supporting tree more fruitful, and that it does but little harm. The idea is a very mistaken one, for Shakespeare truly terms it "the baleful mistletoe." The parasite may and often does throw the tree into bearing. The tree seems to make an effort, as it were, with the knowledge that it is attacked by a vital enemy, which will never leave until it has completely destroyed it, branch by branch. The tree in a few years begins to shrivel and decay, and the fruit grows smaller year by year, albeit the tree may keep up the struggle for many years.

Something may be done to help the trees. The Mistletoe should be attacked bodily, and all established plants broken off, or cut closely year by year. If this is done before Christmas, the berried branches will readily sell at any Railway Station at £4 the ton. The only effectual remedy however is to destroy the seeds, or seedlings. The silvery seeds are deposited by the birds on the branches, and the first rain washes them to the underside, where the glutinous matter causes them to stick. Here the birds are useful. Tits and Finches happily eat many of them, and the quick eye of the Orchardist should enable him to destroy many more with his spud. If the seed is left alone, the young Mistletoe seedling will send its root down the inner bark, and throw out its first leaves the second or third year. Nothing now can be done but to cut off the branch, if the Mistletoe is distant from the trunk; or if not, to check its growth, by constantly breaking it off. When the tree gets thoroughly affected, its place should be supplied by another from the nursery.

CANKER.—This disease, the terror of all Orchardists, and the bane of most Orchards, is due to debility, which may arise from many causes. Canker is almost always actually caused by direct injury from accident, or sudden variations of temperature. In all cases there is a want of vitality. The tree is old, or delicate. The soil is not sufficiently drained, or it is too poor, or for some cause, does not suit the variety. Any direct injury to the bark of the tree will frequently give rise to Canker, whether this is produced by the accidents of wind, or ladders, or clothes lines, or friction of one bough or another, or by sudden alterations of temperature (the severe frosts of winter, acting on insufficiently ripened wood), or by a sudden check to growth from a late frost in spring; all which causes lacerate the vessels of the young wood, and Canker appears in the following summer.

Canker commences with enlargement of the vessels of the bark more apparent by the way in the Apple, than the Pear tree. It continues to increase until in the course of a year or two the *Alburnum* dies, the bark cracks, rises in large scales, and falls off, leaving the stem dead and ready to break off with the first wind, if it be not before removed.

The best treatment of Canker is to remove the parts injured and give a good supply of nourishment to the affected trees. "Want of food" said a good orchardist, "I have always found to be the cause of Canker, and the same may also be said of the woolly Aphis. My young trees in the hedge rows became badly effected with Canker, and it occurred to me, that the thorns took the nutriment from their roots. I fed them with a dressing of lime, cowdung, and fresh mould, on the surface of the ground. This soon produced a good effect, and the trees recovered their luxuriance. I have never let my trees want food since and am always rewarded by their healthy condition and abundant crops."

INSECT BLIGHTS.—The great German Entomologist, Kaltenbach, gives the number of different Insects that attack Apple trees as 183. Of these 20 are Aphides, 32 are Beetles, 115 are Moths and Butterflies, and the remaining 16 belong to other classes. It is only necessary to mention a few of the more destructive ones. The apple Weevil *(Authonomus pomorum)* is very destructive to the blossom and is often very abundant. The *Aphis mali* attacks the young foliage and setting blossoms, and its ravages are often too widely spread to admit of any direct effective remedy. The apple grub, *carpocapsa pomonana* attacks the fruit in all its stages.

The most destructive perhaps of all Insect Blights to the trees themselves, is the AMERICAN BLIGHT *(Aphis lanigera)*. It attacks the woody part of the tree and is very fatal. The insect attaches itself to any part of the tree where the cuticle is broken. It is viviporous like most other *Aphides*. It lives on the sap of the tree, and by its irritating presence causes excrescences of growth, and eventually, the death of the branches beyond its situation. It is the habit of this *Aphis* to retire into the ground during the winter, and cluster in the crevices of any available roots, and here it may be advantageously attacked. The first remedy again, is to feed the tree, then to get at the pest directly, as far as possible, by applying a weak mixture of petroleum with soft soap, say, an ounce of petroleum and half-a-pound of soft soap, boiled gradually in a gallon of water, and apply with a brush wherever the woolly insect shews itself. This remedy has the additional advantage of attacking its winter quarters at the foot of the tree, as it is washed there by the rain. This petroleum emulsion is very troublesome to keep well mixed, and when the blight is not very extensive, a moderately strong solution of soft soap, or of agricultural salt, is much more easy of application, and often very effective.

RED SPIDER. *(Gamasus telarius)*. This insect is sometimes very destructive to the leaves of Apple and Pear trees. It is believed to be due to the state of the soil in which the tree grows, which may be too light, or too poor for it, and this belief points out the direction in which relief must be sought.

Many other insects attack Apple and Pear trees, such as *Episema cæruliocephala; Cheimatobia brumata; Porthesia auriflua; Lozotænia rosana; Tortrix heparana; Tortrix ribeana; Tinea corticella; Curculio vastator; Semasia Wæberana;* with several other species of *Aphis, Acarus,* and *Coccus.* The visits of these enemies however are for the most part local, and their presence can only be met by the partial remedy of smoking to windward, where there may be plenty of damp straw or mouldy hay, at hand, to give the opportunity of doing so.

FUNGUS BLIGHTS.—Fungus growths are always unwelcome guests in an Orchard. A botanist may admire a fine *Polyporus hispida*, or rejoice in a magnificent cluster of *Agaricus (Pholiota) squarrosus* with its leopard-like spots growing from the bole, or at the foot of an apple tree, as it so commonly does; but these with all their tribe do but indicate decay within. They should of course be removed at once, though the disease upon which they have fed will exist there still.

MILDEW.—Blight, or Mildew, is another fungus, or Microscopic *Oidium* growing on the young leaves and shoots of the tree. It may appear at any time from Spring to Autumn. It causes first a white mealy appearance of the young shoots and leaves; which then curl up, grow black, and drop off to the great detriment of the trees, if the Mildew has attacked them extensively. This fungus appears under certain atmospheric conditions, such as moisture with a sudden prevalence of cold winds, checking growth. Its remedy is well known to be sulphur, when it admits of proper application, which can seldom be the case in an Orchard. The common practice of whitewashing the trunks of the trees would render good service in checking all Fungus Blights, if they would but remember to add to every gallon of whitewash a handful of sulphur to be exhaled by the sun, during the heat of summer; and if too they add a handful of soot to sober down the glaring effect of the whitewash, it will be a great improvement.

RUST *(Helminthosporum pyrorum).*—This is another microscopic Fungus, which, in cold wet summers, as in that of 1879, is most destructive to the Perry Orchard. It appears in patches on the leaves of the Pear trees and on the fruit, and seldom ceases to spread, so long as a leaf, or a Pear, is left on the tree. *Rœstelia cancellata* and some other microscopic plants could also be named, but their presence and power of inflicting injury, depend more on the season, than on any other cause; and they admit of no remedy that would be available over the extent of an Orchard.

VEGETABLE BLIGHTS.—Lichens and Mosses of many kinds often abound in low lying Orchards. The Lichens, commonly called "Old Man's Beard" *(Romalina fastigiata, Evernia prunastri, Usnea barbata, &c.),* sometimes completely cover the great and small branches of the trees. They are the attendants of a damp atmosphere, and derive their sustenance from it. The trees want more air and sunshine, and the ground is seldom well drained, where the trees present this aspect. The only injuries they occasion the trees is by preventing the access of light and air to the branches, besides harbouring the numerous leaf eating and other insects, whose presence is not desirable. Drainage, pruning and feeding the trees, are the best preventive means. When the Mosses and Lichens exist on the trees of high value, it will perhaps repay the trouble to scrape them off and wash the boughs with a strong solution of soft soap, or with lime water, to check their fresh growth.

OTHER TREE ENEMIES.—All orchard writers dwell at considerable length on many Orchard enemies, such as Cattle, Hares, Coneys, Moles, Water-rats, Birds, Snails, Caterpillars, Pismires and Ants. These enemies must be met, as they occur, by the practical ingenuity of the Orchardist. The most real are Hares and Rabbits, which in severe weather, when the ground is covered by snow, and other food is scarce, will sometimes destroy an orchard by barking the young trees. In addition to the gun, the best remedy is the lime and sulphur wash, very freely applied. Furze if at hand may be tied round the tree stems; but wire netting is the only effectual remedy, when the animals abound and their need for food is pressing. The use of grease, tar, or petroleum, so commonly recommended, are better avoided, since they are apt themselves to be injurious to the young trees.

III. FRUIT MANAGEMENT.

The customs which prevailed in the Orchard two hundred years since are very different from those followed at the present time. The early Ciderists divided their fruit into three classes. The first consisted of such Apples as would make a summer cider for immediate drinking; as the *Codlings; Jenettings; Spice Apple; Summer Queening;* and all the early summer fruits. The second class consisted of those that made the best, and richest, and longest keeping cider, and embraced all the established varieties of cider fruits, as *Gennet Moyle; Redstreak; Bromsberrow Crab; Golden*

Pippin; *Westbury Crab*; *Harvey Apple*; *John Apple*; *Underleaf*; *Stocken Apple*; *Oaken Pin*; *Elliot*; *Nonsuch*; *Musts, Fillets*, &c., &c. Lastly, the third class contained such fruits as made "a pleasant, sweet, acceptable Cyder, though not long lasting" useful for the table, such for example, as *Pippins*; *Pearmains*; *Gillyflowers*; *Marigolds*; *Golden Rennetting*; *Winter Queening*, &c. The early Ciderists thus recognized the fact, that in Cider districts, Cider could be made from all varieties of Apples; but at the same time they shewed a keen appreciation of the varying qualities of Cider, made from different varieties of fruit. In the present days of cheap and easy transit, the first and third of these classes find a more lucrative sale in the markets for domestic consumption, and they are only used for making Cider in some exceptional year, or for some peculiar reason. The Apples now used for making the best quality of Cider, and the same may be said of Pears for Perry, are special varieties grown for the purpose; and such as are not worthy of consideration for use in any other way. They vary as a matter of course as to their season of maturation, and are therefore practically divided into early, mid-season, and late varieties; and thus in well regulated Orchards, the mill is supplied in convenient succession. In the Channel Islands, in Germany, and sometimes in America, it is still the custom to use the best varieties of dessert fruit, both of Apples and Pears, for the manufacture of Cider and Perry, but it can scarcely be said that the result justifies the practice.

FRUIT GATHERING.—The first care of the Orchardist is to gather the fruit when sufficiently ripened, and this period will vary considerably, not only according to the season, but also according to the varying aspects of the tree.

> "Fruit gathered too timelie will taste of the wood,
> Will shrink and be bitter, and seldome prove good;
> So fruit that is shaken, or beat off a tree,
> With bruising or falling soon faultie will be."
>
> TUSSER.—*Points of Good Husbandry*.

The ripeness of the fruit is generally indicated by the change of colour, by the perfume and flavour of the fruit itself, by the blackness of the pips, and by the fact of its beginning to fall from the tree; but the experience of the fruit grower enables him easily to recognize the proper time for gathering it, even in the varieties in which these signs may not be very manifest. The earlier kind of Pears and also of Apples, will generally be ready about the end of September, and with this early fruit it is customary to mix such of the windfalls as may be in good condition; and thus clear the ground for the better qualities of fruit. The gatherings from which the best cider is made usually occur about the second or third week of October, and by the end of the month the trees should be cleared of even the latest varieties.

> "The moon in the wane gather fruit for to last,
> But winter fruit gather, when Michael is past."
>
> TUSSER.

The mode of gathering the fruit also demands attention. The better kinds of fruit, such as are required for the market or domestic use, must be carefully hand picked, since every bruise will injure them; but this extreme care is not necessary for the varieties required for Cider or Perry now under consideration. These may be gently shaken from the trees on to a layer of straw, unless the grass is abundant. A coarse cloth, or piece of sacking, placed to receive the Apples is very convenient for removing them. The simple plan recommended by Marshall cannot be surpassed. As soon as the spontaneous fall of fruit begins to take place he recommends the first gathering to begin. The boughs should be gently shaken by means of a pole with a hook attached to it, but the fruit that sticks firmly to the tree must be left to become more mature, and be shaken off at a later period. This practice is still followed in the best Orchards, when the trees are thus gone over, three, or sometimes four times, at intervals of ten or twelve days, until the whole crop has been matured and collected. The fruit which falls the second time, is considered the most favourable for the best and strongest liquor required for bottling.

APPLE HEAPS.

"The farmer, with foreseeing view,
Prepares himself for the forthcoming spring;
Nudg'd by the ripen'd fruit that silent falls
On the long grass beneath; at early morn
He clears the Orchard boughs, and piles the fruit,
And the press gushes in the pleasant juice."

PARTRIDGE'S *English Monthly*.

As the fruit is collected from the trees it is placed in heaps, until it becomes ripe and mellow for the mill. There has been much discussion as to the position and formation of the Apple heaps. The common practice is to place them on the plain ground in the Orchard itself, or in some convenient place by the homesteads. They are usually made from about eighteen inches to two feet six inches in thickness, and left without any protection either from the sun, from the rain, or from frost, not to mention the fowls and wild birds. Thus they remain for some two or three weeks, to as many months, with the later varieties, to suit the convenience of the cider maker. Marshall recommends that the fruit after being collected perfectly dry should be laid up under cover in an open shed, or where a thorough current of air can be had, in heaps not more than ten inches deep. The best writers of the 17th century gave the same advice. Marshall admits that this practice was not followed in his day any more than it is at present. It is scarcely likely ever to be followed in large Orchards, although the advice is both good and sound.

The different varieties of Apples should always be placed in separate heaps, so as to insure their being sent to the mill when uniformly ripe and mellow. With the exception of a few varieties of noted excellence, Cider is made from different sorts of Apples mixed together, and here the good judgment of the Cyderist comes into play, to mix the varieties which will best improve each other.

"There are, that a compounded Fluid drain,
From different Mixture, *Woodstock*, *Pippin*, *Moyle*,
Rough *Elliot*, sweet *Pearmain*, the blendid streams
(Each mutually correcting each) create
A pleasurable Medly, of what Taste
Hardly distinguish'd." PHILLIPS, *Cyder*.

Apple heaps on the open ground may be from one to two feet in thickness, without fear of the fruit becoming heated, but on a dry floor, the depth should not exceed one foot. The heaps should most certainly be protected from changes of weather, and if placed in rows this may readily be done by thatched hurdles resting on a pole running above the heap. These may be easily moved or replaced, and if frost should set in be covered with clothes or tarpauling. The sun, by causing a partial fermentation, is injurious to the apple heaps, but still more is the rain. If this is doubted, let a whole and sound Apple be placed in a glass of clear water and allowed to remain there for seven or eight hours. In this short time, the water will have taken a rosy hue, and the sweet taste of the Apple, whilst the Apple itself will have lost much of its flavour. The explanation is, that by the natural laws always in operation, between fluids of different density, the law of endosmore and exosmore, the water has passed into the Apple, and juice has passed out into the water, greatly to the injury of the fruit. Frost is also very injurious to fruit, for when it has once been frozen, it will never ferment properly. A French chemist found the loss to be about one and a half per cent. of alcohol from the juice of fruit that had been frozen.

It is most desirable therefore that fruit should be protected, but it is seldom done. During the Autumn and Winter of the years 1878-9, and 1879-80, though fruit was scarce, rains frequent, and frost severe, it was a rare circumstance to see the Apple heap protected.

Cider makers in all ages have agreed, that the fruit must be used in its best condition to make good Cider. As it is put in the baskets at the heap, to be carried to the mill, all the bad Apples should be rejected. Unripe fruit contains but little sugar, or flavour; heated, or frost bitten fruit will not ferment properly; and bruised and rotten fruit introduce elements of injurious fermentation. All watery or inferior fruit should be ground by itself, since it must of necessity make inferior liquor. To use all the fruit together, ripe and unripe, good and bad, is fatal to the production of superior Cider. The quaint remark of Worlidge on this careless

custom is as true at this time, as it was in his own days (1675).
"This error, or neglect, hath not onely been the occasion of much
thin, raw, phlegmatical, soure, and unwholesome Cider, but hath
cast a reflection on the good report that Cider well made, most
richly deserves," and he adds very sensibly, "better lose part of the
Cider, than spoil the whole."

Pears are not considered to require so much care and good
management as Apples do before they are carried to the mill, and
the usual custom is when the fruit begins to fall freely, to shake off
the remainder of the crop, and grind the whole without delay.
The long keeping varieties require to be placed in heaps as Apples
are, and are of course much improved by being allowed to become
uniformly ripe.

THE MILL.

> "Lo! for Thee my Mill
> Now grinds choice Apples, and the British Vats
> O'erflow with generous Cider."
>
> PHILLIPS " *To his friend Harcourt in Italy.*"

The mode of extracting the juice from Apples and Pears
to make fermented liquors, seems to have been of the rudest
kind until a comparatively recent period. The fruit was
grated or crushed in any rough and simple way, and since the
quantity made was but trifling and labour cheap, it answered
sufficiently well. Worlidge writing near the end of the 17th
century says "The operators did beat their fruit in a trough of
wood or stone, with beaters like unto wooden pestles, with long
handles, whereby three or four labourers might beat twenty or thirty
bushels in a day." When a large quantity of fruit was grown and
Cider and Perry became articles of commerce, it was necessary to
find out some process more economical and expeditious. The
happy idea occurred to some one—whose name is lost to a grateful
country—to make the trough of a circular shape, and roll round a
heavy cylinder in it. The original mill was of rude construction,
and both the wheels, or cylinder, and trough, were made of wood

studded with hobnails. The wooden cylinder soon gave place to stone for the advantage of its weight, and this entailed the necessity of making the trough of the same material. A mill thus constructed and worked with one horse, crushed the fruit so rapidly as to make from two to three hogsheads daily.

> "Blind Bayard, worn with work and years,
> Shall roll the unweilding stone from morn to eve."
>
> PHILIPS, Cyder.

Dr. Beale, in Evelyn's "Pomona," speaks of some mills so large as to be able to grind half a hogshead at a time. The construction of such a mill required the heaviest and most durable stone. In Herefordshire, the Millstone Grit from the Forest of Dean, soon came to be noted as best suited for this purpose. Such a mill, was necessarily expensive, and so efficient, that at first one mill would serve for the district; the grist in the shape of Apples and Pears being brought to it from all the surrounding Orchards. In course of time, as Orchards multiplied, every large farm had its own mill, and at the present time they are very numerous and most of them regularly used. The great fault of the stone mill is, that the pulp is apt to roll too quickly before it, and the fruit may thus escape being evenly crushed, a fault not altogether obviated by the diagonal grooving. The trouble of removing the pulp from the trough is another of its disadvantages.

About the year 1689, Worlidge invented a moveable iron mill, which he called the "Ingenio," a name borrowed from the Cubans, who curiously enough grind their sugar canes at the present day, with a machine thus called. With this mill he tells us, that "two labourers, one feeding, and the other grinding, can manage eight bushels an hour by interchanging all the day, with ease and delight." The "Ingenio" was introduced into Somersetshire many years before it reached Devonshire, Gloucestershire, or Herefordshire, where the Stone Mills were in general use. Marshall speaks of the Stone Mill as "an unfinished machine"—whilst Thomas Andrew Knight, some twenty years later, attributes much of the celebrity of Herefordshire Cider to the perfection of the Stone Mill; and this feeling in its favour exists more or less throughout the county at the present time.

The French have paid great attention to their fruit machines. They have one, the "Écraseur," (Salmon and Bergot), which grinds seventy-five bushels per hour with ease; and this has now been surpassed by the "Écraseur Universal," which with only one pair of granite cylinders will grind two hundred bushels of fruit per hour, besides being ready at other times to do the whole work of pulping the roots of all kinds, which may be required on the farm.

Various mills have been invented of late years. Mr. Davis, of Linton, near Ross, has introduced an admirable machine, in which the crushing power, by a clever application of the French principle, is very considerably increased by causing the two stone cylinders to rotate at different degrees of speed. Indeed there is some fear of the machinery becoming too rapid and too perfect to obtain good Cider.

A traction steam engine in these days draws the mill and an attendant press into the Orchard; grinds up the fruit heaps at a rapid rate; and presses the pulp forthwith. The math, or cake, is rejected on the spot, and the casks at once filled with the must. The whole process is completed, with an economy of time and labour that can scarcely be exceeded. The economy is false, when the result is taken into consideration, for the best Cider is not to be made in this way. If the mill were taken from time to time to the Orchard as the different varieties of fruit ripened, the economy would be lost. And thus it comes to pass that all the Apples are ground up at once—early and late varieties—ripe and unripe—they are all submitted together to the mill and the press, No time is allowed for the pulp, or "pommage," as the old writers call it, to commence fermentation exposed to the air, or for the juice set free to extract the full flavour of the fruit from the rind, the pips, and the more solid parts, and thus the liquor loses flavour, and the so-called economy defeats itself.

GRINDING.—The degree of fineness to which the fruit should be reduced into pulp has been much discussed. The old writers considered the fruit need not be ground very small, though it was the common practice in their day to do so with the view of getting more juice. Marshall says that in the South and everywhere except in the Cider Counties, it was believed that the cellular juice of the fruit alone formed the necessary ingredients of good Cider. In Herefordshire it was always commonly believed as it may be said to be now, that the flavour of the Cider was chiefly derived from the kernels, or pips, and the colour from the skin of the fruit; and it was therefore held to be all important that the pips should be crushed in the mill. M. Berjot, a distinguished French chemist, who studied the subject closely, and who wrote a prize essay on the "Chemical Analysis of the Seeds of Apples," proves by numerous experiments, that for Cider of the best quality, it was better not to crush the pips, because the diffusible odour of the essential oil they contain, spoilt the delicate flavour of the Cider; but with fruit of an inferior quality, deficient in flavour, it was an advantage to do so, since the pips gave their own flavour to it, and took away the earthy taste it is otherwise apt to have. M. Berjot invented a mill specially designed to tear up and crush the fruit without bruising the seeds.

Monsieur Hauchecorne distilled the spirit from Cider made with the pips, and from that made without pips, and obtained excellent brandy from both, though the flavour was different. The judges pronounced them to be equally good.—"*Le Cidre*," p. 341.

The common belief, therefore, that it is necessary to crush the pips to obtain the best quality of Cider, is not correct; and the impression also, that its colour is derived from the skin is equally wrong, for, as was pointed out by Marshall, the palest coloured Apples often produced the ruddiest Cider. He instanced the *Hagloe Crab*, and it is equally observable in Cider from the *White Must*, the *Forest Styre*, the *White Bache*, the *Yellow Hereford*, and several other Apples, that have but little, or no colour themselves.

In grinding the first portions of fruit, especially in a dry season, it is necessary to sprinkle water over the Apples, " to wet the mill," as it is termed. The juice first procured will be used to give moisture to the succeeding grindings. The facility with which water may be added is to be lamented, for in this way the character of the Cider is often much deteriorated. It may be sold at a cheaper price, but increase of bulk increases trouble, and therefore expense, and the adulteration prevents the possibility of obtaining the price which a better cider would command.

The solid portions of the pommage, that which remains in the pressing bags, now called the "math," "cake," or "cheese," and by old writers "powz" or "murc," is often re-ground at the mill, with the inferior fruit, and the addition of a considerable quantity of water. In this way an inferior Cider, Cyderkin, or Purre, for home use, is legitimately made.

> "Some when the Press by utmost Vigour screw'd,
> Has drained the pulpous Mass, regale their swine
> With the dry Refuse; thou, more wise, shalt steep
> Thy Husks in Water, and again employ
> The pondrous Engine."
>
> PHILLIPS, *Cyder*.

The math, or cake, is sometimes mixed with chaff, and given to the cattle; in small holdings, when dry, it is used as fuel; or lastly it is thrown on the special manure heap to be returned in this way to the Orchards.

By common consent Pears require comparatively but little grinding.

IV. FERMENTATION.

CLEANLINESS.—In all the varied processes for converting the juice of the Apple and Pear into Cider and Perry, from the very beginning to the end, the most scrupulous cleanliness is required. The Mill should be thoroughly cleaned before the fruit is brought to it; if of stone it must be scrubbed throughout; the iron clamps which unite the stones, and especially the leads, which fix them, must be carefully cleaned. If it is an Iron Mill, not only should the stone rollers be scoured, but any rust that may have formed on the framework should be rubbed off. The juice of the Apple will not dissolve the metal, unless it is left long in contact with it, but it readily dissolves the dull grey powder which forms on lead, or the brown rust of iron, which are oxides of the metals; while the salt formed, being soluble, is carried through the process of fermentation, and remains in the Cider. The salt of iron, if strong enough to be injurious, would discolour and spoil the Cider, so no more need be said of it; but the salt of lead is more dangerous, since it sweetens the Cider, and gives no evident sign of its presence.

In almost all the Cider districts, the most painful cases of colic frequently occur from want of care to prevent the contact of apple juice with lead. Sometimes a portion of juice is left in the trough of the mill for many days, which dissolves the oxide on the lead soldering of the clamps connecting the stones together; or sometimes when white lead has been most thoughtlessly used by the cooper, as caulking to prevent leakage in the casks, it is present in its most dangerous form, for the juice dissolves it easily.

> "Evil is wrought by want of thought,
> As well as want of heart."
>
> HOOD, "*The Lady's Dream.*"

Or lastly, and this is perhaps the most frequent cause of all, the Cider takes up the lead from the Cider engine at the bar tap; and the "Boots," who drinks the first jug drawn in the morning, instead of throwing it away, as directed, as is the general rule, gets a most painful and serious illness. TOO MUCH CARE CANNOT BE TAKEN TO PREVENT THE CONTACT OF CIDER WITH LEAD, EITHER IN ITS MANUFACTURE, OR IN ITS PRESERVATION.

The barrels or casks must always be carefully examined, and if not perfectly clean and sweet they must be made so. Scalding with boiling water is the common practice, and some first press through the bung hole a yard, or more, of stout iron chain, with a cord attached to one end for its removal, and roll the barrel about well. A powerful jet of steam thrown into the barrels is far more effective than boiling water, for obvious reasons, where circumstances admit of its application. If not then perfectly sweet, sulphur should be burnt in the barrel, and the scalding be afterwards used again. It is far better, however, to take out the head of every foul, or even doubtful cask, that the cleansing may be thorough and effectual. This last excellent practice is followed by some of the best Cider makers as regularly as the season comes round.

The Cider house, wherever it admits of it, should be closed as tightly as possible, with all the vessels and implements used there, and then freely fumigated by burning sulphur within it. In this way any germs of injurious fermentation that may exist there from previous operations would be effectively destroyed.

CHEMICAL COMPOSITION OF FRESH APPLE JUICE.—The chemists appointed by the French Congress for the Study of Cider Fruits, have given the following analysis of fresh Apple Juice, as the mean of many examinations of Juice from the best varieties of fruit; their density varying from 1067 to 1080.

1000 Parts of Juice contained of:

Water...	800
Sugar capable of being converted to Alcohol	173
Tannic Acid, or Tannin	5
Mucilage, or Pectos'ne (soluble Pectine Gum)	12
Free Acids (Malic, Tartaric, &c.)	1·07
Albumen and Fermentable Matter	5
Saline Matters (Lime, Mallates of Potash and Lime, Phosphate of Lime)	1·25
Pectic Acid, Coloring Matter, Fixed and Volatile Oils, and Insoluble Substance in Suspension)	2·18

They found the juice from inferior fruits contained the same materials, but in very different proportions, with the exception of Albumen, Fermentable Matter, and the Salt of Potash and Lime, which were in much the same proportions in all qualities of juice. These inferior juices, having a low density, had one third less of

Sugar, the Tannin was only 1 part, instead of from 4 to 6; and the amount of Mucilage was only 4 instead of 12 parts in a thousand. "*Le Cidre*," p. 3.

From an examination and comparison of the best Cider fruits of France, England, Germany, and America, the Congress states that the general characters they possess are, good Perfume, slight Bitterness, and very little Acidity, with a notable quantity of Tannin, and of Mucilage, and a very large amount of Sugar. Sugar, and the Alcohol formed from it, are the most important elements of Cider, and the best varieties of fruit are essentially necessary for their production in sufficient quantity to enable it to travel without injury. The best and soundest Cider should contain from 8 to 10 per cent. of Alcohol (the French say 12 per cent.); with from 2 to 3 per cent. of sugar remaining still unreduced, to give it the highest commercial value.

M. Pasteur gives, as the result of many analyses, that 100 parts of the Sugar of Fruits capable of being converted into Alcohol yield:

Carbonic Acid	46.67
Alcohol	48.46
Glycerine	3.23
Succinic Acid	0.61
Matter yielded to the ferment	1.03

Tannin, or *Tannic Acid*, is the next most important element in the fresh juice. It makes the liquor "fine" more readily, by causing the Albumen, the Pectine, and the Yeast plants to be deposited; and thus acts indirectly as an antiseptic, regulates the action of the fermentation and prevents the after tendency to ropiness, so apt to appear in the liquor from fruits of great richness. It is believed also to possess the great virtue, by its astringent qualities, of moderating the action of the Alcohol on the system, in the wine and other liquors containing it; and thus rendering them at once less exciting, and more strengthening. The French chemists state that it requires from 2 to 3 parts in 1000 to produce its full effect in the manufacture of Cider and Perry, and that from 2 to 3 more in 1000 should be present for its wholesome qualities.

The *Mucilage*, or *Pectosine*, when abundant, forms another element of distinction between good and bad juice. It renders the fresh juice more thick and viscous, and eventually gives softness and body to the liquor. It helps also to preserve the Alcohol, by opposing the Acetous fermentation, and is thus always present in long keeping Cider and Perry.

The *Malic* and *Tartaric Acids* give the refreshing character to Cider and Perry, which is so desirable in a summer beverage. The Malic Acid abounds most in Apples, and Tartaric Acid in Pears, and their too great abundance is rather to be feared, than their failure. The acidity these acids give, together with the *Perfume* and *Bitter Principle* in the juices, which tend also to render the Cider and Perry more pleasant and agreeable, are difficult to to determine chemically; but their proper quantity can be estimated with sufficient practical accuracy by smell and taste. An excess of acidity—is the chief characteristic in fruits of poor character.

THE THEORY OF FERMENTATION.—The natural saccharine juices of all fleshy fruits, if left to themselves at a temperature of 50 to 80 degrees, will immediately begin to take on vinous fermentation, especially if they are acid, which is usually the case. This fermentation, moreover, takes place without the addition of any substance to bring it on. Thus, if pulp, or pomage, from ripe apples, or pears, as taken from the mill, be left to itself at the ordinary autumnal temperature, minute bubbles are soon observed to rise to the surface and form a white froth; its bulk will be increased, and if the thermometer is plunged into it, its temperature will be found to have risen. These changes are due to the commencement of Alcoholic Fermentation. The bubbles contain Carbonic Acid gas, and if the juice is tasted, it will be found to have acquired a spirituous fragrance due to the formation of alcohol. Why should the simple crushing of ripe fruit lead up to a series of changes, so curious, and yet so certain? The distinguished Frenchman, M. Pasteur, has occupied some years of his life in attempting to answer this question. He has concluded a long series of experiments, requiring the utmost patience, with the closest attention to minute details, and he happily possesses the genius, which has enabled him to arrive at many very interesting and important results.

M. Pasteur has succeeded in proving that on the external surface of all fleshy fruits when they become ripe, there exists certain minute particles, or germs, which when brought into contact with the ripe juices of the fruits, develop into minute plants, and forthwith grow with great rapidity. These plants are the Yeast Plants, which belong to a great family of microscopic funguses. They are called *Saccharomyces*, or Sugar-Eating Funguses, from the peculiar power they possess of decomposing and living upon the saccharine principle of plants, the grape sugar, or Glucose, as it is termed by the chemists, and thus causing their elements to be rearranged into Alcohol, Carbonic Acid Gas, Glycerine, Succinic Acid, Volatile Acids and other products.

M. Pasteur obtained these corpuscules, or germs, by washing ripe fruit—grapes he first used—with pure distilled water. The water was rendered slightly turbid by the presence of an infinite variety of minute particles. Under the microscope, many of them were shapeless atoms of dust, scales of epidermis, or spicules of crystaline matter, but many others appeared to be organised corpuscules, resembling the spores of funguses. These organised corpuscules differed considerably from each other, and when M. Pasteur cultivated them with all due care in saccharine fluids, he found them to swell and germinate at different times and in different ways. In an hour, and often in less time, he observed a copious formation of new cells, whilst small bubbles of Carbonic Acid gas were given off, showing that the formation of Alcohol had already begun. They were thus proved to be true Yeast Plants, or *Saccharomyces*. M. Pasteur traced the growth of several species of Yeast Plants under the microscope, all differing in their size of cells, shape, mode of budding, and general growth. The most common of these plants, to whose growth in the natural saccharine juices of ripe fruits the formation of Alcohol is chiefly due, he described minutely in 1862, in the *Bulletin de la Société Chimique*, p. 67. These observations were fully confirmed by Dr. Rees, a

German physician and naturalist, in 1870, and he first attached to them, the following specific names:

Saccharomyces apiculatis; which is the first to grow, and the most minute in size.

Saccharomyces Pastorianus; which is by far the most active and abundant, and which Dr. Rees named after M. Pasteur: and lastly,

Saccharomyces ellipsoides; which is the slowest in growth, but most persistent, and which forms the ordinary ferment of wine.

M. Pasteur describes minutely from his observations, the life history of these *Saccharomyces*, their several modes of rapid development and reproduction, together with the chemical changes they effect by the decomposition of Glucose, such as the production of Alcohol, Glycerine, &c. These plants frequently take different forms, according to the varying circumstances under which they grow; for example, one form of *Saccharomyces Pastorianus* is so small that it was at first thought to be a different plant, and it was called by Dr. Rees *Saccharomyces exiguus;* whilst another form was named by M. de Bary *Dematium pullulans.*

Certain it is that the mode of life of these plants is essentially different from that of all other living organisms, and the resulting chemical action is equally exceptional. Most organized beings live and grow by absorbing Oxygen from the air, and setting free Carbonic Acid: so do the *Saccharomyces* in the first stage of their existence; but the air of the fluid in which they live is quickly exhausted, and when this happens, they obtain the oxygen essential to their growth, from the Glucose; and in decomposing the Glucose they set free more Oxygen than they require; and this uniting with the Hydrogen and Carbon present, forms the various products of the fermentation they occasion.

There are numerous other microscopic funguses whose minute germs are always present in the air, ready to take their life-growth in the decomposition of saccharine fluids; such as various species from the families *Mucedines, Mucorina, Torulæ,* &c. The fermentations these plants produce, are commonly called "after fermentations." M. Pasteur calls them "diseased," because their propagation and development is always attended with the loss of Sugar, or Alcohol, and also with the production of some unpalatable result. To apply this rule to our present subject, the Cider or Perry may thus become acid, viscous (ropy), or be altogether spoilt, according as the germ of the several funguses, which produces these results, have been able to develop themselves within it.

ACETIC FERMENTATION.—The fungus which causes this dreaded fermentation is the *Mycoderma Aceti.* Its germs are so minute as to be only perceptible with a powerful microscope when they are diffused in the liquor; but when aided by exposure to air under a high temperature, they are quickly developed into chains and chaplets, which massed together, soon appear as a film of grey mould floating on the surface, and this is commonly called "flowers of vinegar." When this film has grown thicker and become submerged, it takes on a gelatinous form of surprising toughness and lubricity, and it is then called the "Mother of Vinegar," or the "Vinegar Plant." The *Mycoderma Aceti* requires a warm temperature, and a much more abundant supply of air, than do those *Saccharomyces,* which cause Alcoholic fermentation, and the more freely air is supplied the more rapidly the plant grows, and the more quickly is the vinegar produced. The *Mycoderma Aceti* has the power of decomposing Sugar, or Alcohol, either singly or in combination, producing Acetic Acid and water, without the evolution of Carbonic Acid gas. When the access of air is prevented, as should always be the first care of the Wine or Cider maker, its action is extremely slow. It is sure, nevertheless; for the germs that find their way with the liquor into closed vessels and well-corked bottles will prevail in the long run; thus an excellent bottle of Wine or Cider will end in becoming a bottle of Vinegar, though it may take half a century to effect the change.

VISCOUS FERMENTATION, or ROPINESS.—This "disease" is also caused by the rapid growth of the minute spherical germs of a fungus, not as yet specifically named. It quickly develops itself into chains of vesicles, and in this process changes the Glucose of liquid into Gum and Mannite, with the evolution of Carbonic Acid gas. In some seasons Ropiness is very troublesome, and remedies in abundance have been recommended to check it, in accordance with the prevailing belief on the spot, as to its cause.

PUTRID FERMENTATION, it need hardly be said, is not due to the growth of Fungus Plants, but to the presence of *Bacteria*, *Vibriones*, and *Infusoria* in general, whose germs are also always present in the air, and when deposited, under circumstances favourable to their growth, develop themselves with great rapidity to the destruction of the liquor.

M. Pasteur, having proved that the VINOUS FERMENTATION of Saccharine fluids was caused by microscopic plants, growing from germ cells found on the surface of ripe fruit, next endeavoured to account for their presence. Infinitesimal in size as they are, and only perceptible by the aid of the microscope, he concluded that they formed part of the dust wafted about in the air. The germs themselves, and their mode of growth, he found to resemble the spores and habit of growth of certain funguses of the family group of *Dæmatiei*, which are common on dead wood during the Autumn. Some species of the family, there is reason to believe, produce two forms of germ cells, the one set adapted to aërial growth, and the other capable of living when submerged in fluid, by decomposing the substances with which they come in contact. Thus Alcoholic Fermentation may be briefly defined as "A CHEMICAL REACTION RESULTING FROM THE DECOMPOSITION OF GLUCOSE BY THE GROWTH OF CERTAIN CELLULAR FUNGUSES."

These striking results of M. Pasteur's labours have met with general acceptance, and they have completely changed the theories Fermentation formerly believed in. The study, however, must be carried much further, before the minute and complicated changes, which are ever going on during the decomposition of organic substances—acting and re-acting on each other as they do—can be fully understood. It is happy for mankind that, guided by practical experience alone, the results of Fermentation have been rendered available, without the necessity of waiting for Science to give the explanation of the various stages of the process.

THE PRACTICE OF FERMENTATION.

"Come let us live and quaff a cheery bowl,
Let Cyder new wash sorrow from the soul."

GAY. *Fifth Pastoral.*

It is agreed on all sides, that the pomage, or pulp of the fruit, should be removed from the Mill, as soon as the grinding is finished, that is, as soon as the Apples or Pears have been reduced to pulp, but there has been much discussion as to how long it should be allowed to remain before being submitted to the press. The old writers state that the general practice in their times was to press the pomage at once from the Mill, and forthwith fill their barrels from the press, but they are also unanimous in advising that the pomage should be placed in open vessels from twenty-four to forty-eight hours, before the "must" is expressed from it. Thomas Andrew Knight held the same opinion. In America the pomage is allowed to remain in an open vat for this time, or longer, according to the prevailing temperature, and an instance is given in Kenrick's *New American Orchard* (1844), where a Mr. Price won the first prize at Concord, Massachusetts, for Cider made from Apples, whose pomage had been left in the open vats for eight days before being submitted to the press. In Germany, and some parts of Normandy, Professor Schlipf states the pomage is left in open vats from five to twelve days, until Fermentation is well established, and the lees begin to settle, when the liquor is drawn off and the remainder submitted to the press. The French press the pomage at once from the mill, except when very occasionally they use table fruit.

The best practice is to place the pomage direct from the mill into large wooden vessels filled to within a foot or eighteen inches from the top. These open vats should be covered lightly with a cloth, or board, and be allowed to remain untouched for two, three, or four days, if the weather is cool. A gentle fermentation quickly begins, and within a few hours, minute bubbles rise to the top and soon form a white froth there. The Carbonic Acid gas escapes from these bubbles, and as it does so, it spreads over the surface of the pomage to the top of the vat, and thus keeps off the action of the outer air, even if left for several days. The advantages of thus leaving the pomage to ferment are, that the juice, and the Alcohol as it is formed from it, are enabled to extract the full flavour and perfume from the peel and solid portions of the fruit, which are so essentially necessary to give a high character to the Cider. The common practice—becoming unfortunately more common still—of pressing the pomage direct from the mill, is, therefore, disadvantageous.

PRESSING THE POMAGE.

"Yet even this season pleasaunce blithe affords,
Now the squeezed press foams with our Apple hoards."

GAY. *Fifth Pastoral.*

When a sufficient time has elapsed, the pomage is taken from the vats in successive portions, and placed upon close textured rough horsehair cloths, and the ends are folded over. Several of these cloths thus filled are placed over each other, a dozen or more at a time, and are all pressed together. In Devonshire, successive layers of fresh drawn clean straw, or reeds, are often used in the press instead of the horsehair cloths. The press is similar in principle to that used for making cheese, but its machinery of late years has been considerably improved, and the whole process of what is technically called "making the cheese" simplified and accelerated. The pressure should be gradual at first, since the first juice runs turbid, and only the latter portion clear. The juice is at once put into large hogsheads, generally holding from 100 to 115 gallons each in Herefordshire, but in Devonshire 50 gallons invariably. The barrels are not quite filled up, a slight "ullage," as the unfilled space at the top of the barrel is termed, being left.

The barrels are placed in a draughty outside building, or a cool cellar, to undergo the most active stage of fermentation. If the temperature is favourable, ranging from 60° to 70°, very evident signs of increased action will soon appear. The bubbles of Carbonic Acid Gas begin to rise so quickly that a constant hissing noise is heard. These bubbles carry up with them to the surface many of the lighter particles of the cellular tissues of the fruit that have passed through the press with the juice, and thus a thick scum is formed on the surface, to which the cells of the yeast plants are gradually added in considerable quantity. This scum soon becomes a thick spongy crust, sometimes called the "upper lees," and it is supported on the surface by the Carbonic Acid Gas arising beneath it, so long as this gas is generated in sufficient quantity. At the same time that this action is going on, the more solid particles of tissue that have escaped through the press, sink into the fluid, accompanied with a considerable portion of the mucilage, and an abundance of yeast cells. This deposit forms the "lees," or "lower lees," at the bottom of the barrel.

As the Fermentation declines the hissing noise moderates, since less Carbonic Acid Gas is generated; the floating crust gets dry on the surface, cracks, and losing its buoyancy, falls in fragments to increase the amount of the lees below. By this time the liquor will have become moderately clear, or will have "dropped bright" as the phrase goes. It should then be racked off and the temperature kept low. This is the crucial point of the whole process, and requires close observation and care; for any delay at this stage incurs the risk of injurious secondary fermentation.

The clear liquor should then be racked, or run off from the lees into a fresh cask, perfectly clean and sweet, by means of a syphon, so as to prevent any unnecessary exposure to the air. A considerable "ullage" should be left in the barrel, and the bung is usually left open for some days. It is better to close the cask with a bung through which a glass tube passes, one end being open into the "ullage" space, and the other outside end bent down and up again, so as to hold a tablespoonful or two of water in the lower bend; or the outer end of the tube may be simply bent down into a shallow cup of water. The advantages of this bent glass tube are, that if an excess of gas is formed in the barrel, its pressure would force it easily through the tube and water; whilst the outer air would be prevented by the water from passing into the barrel. Where the water is put in the lower bend of the glass tube, the amount of pressure within can be estimated by its pressure on the water in forcing it down.

If at the end of a week the liquor remains quiet, and becomes more clear, an ounce of dissolved isinglass should be added to each hogshead, and the bung permanently closed. The isinglass should be first dissolved slowly in a little of the liquor without heat. This will require two or three days. The barrel should remain untouched, until it is required in the spring months for the bottle, or for customers' casks.

The process of Fermentation thus far should have been conducted at a temperature as uniform as possible. It should never exceed 70°, and it should get below 50°. After the final racking it is very advantageous to keep it below 40°, indeed the barrels should be kept at as low a temperature as convenience admits.

Active Fermentation may be said to cease when the hissing noise is no longer perceptible, but it still continues to go on quietly, and the quantity of alcohol slowly increases, and the sugar decreases in proportion, whilst the liquor becomes more clear and bright, acquires a higher aroma, and additional strength.

CIDER MAKING.

NOTES FROM PRACTICAL EXPERIENCE.—The following Cider makers in Herefordshire, whose names usually appear as successful competitors for the Cider prizes offered by the Herefordshire and other Agricultural Societies, have sent to the Woolhope Club an account of the methods they adopt, from which the following abstracts have been made.

Mr. John Bosley, Lower Lyde, Hereford.—" I select my fruit, let it ripen well, and reject all unripe fruit. I crush the Apples to a fine pulp in the old-fashioned stone mill, breaking kernels and all. The pulp is placed in large open vats for 48 hours or more, until it works up well. The juice is then pressed from the pulp and put in a 100 gallon vat, where it remains until it clears itself. It is then racked into another vat and one quart of fine charcoal added and well stirred several times during 12 hours. The liquor is then passed through bags made specially for the purpose. A shooter (as it is commonly called) is placed under the bags to catch the thick liquor which first comes through them; but as soon as the liquor drops bright (sherry bright) the shooter is removed and the clear liquor is allowed to run into the vat below. The thick liquor from the shooter being pressed again through the bags. The liquor is then racked into a well-prepared hogshead and bunged down, loosely at first, but then tightly. Here it remains until it is wanted in the spring for bottling or draught purposes."

Mr. Joseph Davies, Venn's Green, Marden, Hereford.—" Great care and attention is required in the selection of the best fruits, and a great deal of practical experience in blending the different varieties. In my opinion a mixture of fruit makes the best and richest Cider for keeping in bottle. The different varieties should be kept separate in heaps, and be allowed to ripen well, before being used. They should be well ground until the kernels are broken, and without the addition of any water. The pulp should be pressed through hairs, or cloths, and the juice put into casks. When it has worked up well through the bunghole, which will take from 5 to 10 days, the clear liquor should be racked into a fresh well cleaned cask, and the sediment bagged again and its clear liquor replaced. The casks should now be kept cool, at a temperature from 40° to 50°, and left for about a week. Then the same process of racking should be repeated into a fresh cask, and 2 oz. of the best staple or 4 oz. of common isinglass (previously dissolved gradually in some of the cold liquor) should be added. If it should be necessary, from continued fermentation, in 10 or 12 days the liquor must be racked again, and isinglass used as before, once or even twice more, if required. Great care should be taken in racking, and if there is any sediment it should be run through a hair or cloth placed in the tunpail. Bung down tightly and it will be ready for use the following April. The best Cider is made from November to Christmas."

Mr. William Hill, Lower Eggleton, Ledbury.—"The greatest secret in making good Cider is to select the best Apples. They should be gathered every fortnight and placed in separate heaps for two or three weeks, so that each gathering may become mellow at the same time. I use the third picking of well ripened Apples for my bottling Cider, and take care to reject any unkind, unripe or rotten fruit, which are all apt to cause bad fermentation. The Apples should be ground well with the stone rollers of the new mill which can be set close enough to crush the kernels. The pulp is then placed in tubs and allowed to stand until the next day. It is then put into hair cloths for pressing, and the clear liquor put into the hogshead at once. In about a week it will have thrown up a crust to the top, and when this begins to dry on the surface, the clear liquor should be racked into a fresh cask. When this is quite bright, which it generally is by February, fill the cask full, place a piece of brown paper, with a brick on it, on the bunghole for a few days, before bunging it up tightly for good."

Mr. John Watkins, Pomona Farm, Withington, Hereford.—
"The best Apples should be selected, and every variety kept separate as far as possible when gathered. They should be placed in heaps from 12 to 18 inches thick on gently sloping grounds, that any rain may drain away at once. They would be better protected from rain where circumstances admit of it, which is seldom the case in large orchards. Each variety should be ground separately when quite mellow, rejecting all rotten fruit. It is easier to mix the juices of the several varieties after grinding, so as to determine the right proportions of each in blending them to obtain the highest character and flavour in the Cider. The pulp should be allowed to lie from 12 to 24 hours before being pressed. After being pressed the juice should be put into casks and allowed to remain until the the first fermentation has taken place. The clear liquor should then be carefully racked, and the sediment passed through bags of forfar until quite fine, when it is added to the rest. If secondary fermentation sets in, the liquor must again be racked; and if necessary the same process repeated until it has become quite fine and quiet. If it should not then be perfectly bright and clear, a little dissolved isinglass should be added; but the Cider that fines itself without artificial aid is best. It is then tightly bunged and kept in a cool cellar until required for use.

Cider of the best quality, or for bottling purposes, should always be made without the addition of any water. The cake taken from the cloths after the first pressure, is however allowed to soak, with a little water added, in a vat for 24 hours, and then passed through another mill, pressed again and fermented. It then makes a mild and pleasant Cider for immediate use, or for sale at a cheap rate.

Cider is not injured by being frozen. One of the best casks I ever made was frozen for several weeks before it could be racked. The second prize at the Royal Agricultural Society's Exhibition, at Kilburn, in 1879, was afterwards awarded to it.

The great secrets for making good Cider are to obtain good fruit; use it when quite ripe and sound; leave the crushed pulp exposed to the air for some time before pressing it; watch the fermentation throughout carefully; avoid all possible contact with metal, whether iron, which is most common, or lead, which is by far the most dangerous; and to use the most scrupulous cleanliness from beginning to end."

AMERICAN METHOD OF FERMENTATION.

THE AMERICAN METHOD OF FERMENTATION, as described by Downing, consists in placing the newly filled casks, with their bungs out, either in a cool cellar, or in the open air, and as the scum works out the barrel is kept filled with some of the same "must" kept for this purpose. In two or three days the rising will commonly cease, and then the first fermentation is over. The bung is now closed, and in two or three days driven in firmly, leaving a small vent hole open, and this also should be stopped in a few days. The clear liquor is now racked off by syphon into a clean cask, and if in a few days it is found to remain quiet, a gill of finely powdered charcoal is added to each barrel, when it is closed and left until spring. In March they rack again, and if the Cider is not quite bright, they add three-quarters of an ounce of isinglass, previously dissolved, to each barrel. In a few days it will be fit for bottling, and this may be done at any time up to May.

THE FRENCH METHOD OF FERMENTATION is as follows:— The "must" is removed at once from the press into large oak casks well cleaned and prepared for it. They are filled to within three or four inches of the brim, and placed in rows in a cellar with a minimum temperature of 12° centigrade (or 53° Fahrenheit). If the fermentation is slow, they increase the heat to 25° centigrade (or 77° Fahrenheit), by movable stoves. When active effervesence begins to subside and the Cider is "between the two lees," the density of the fluid will be found to have decreased from 1067 to 1035. This is the proper time to rack it, which they do by syphon into casks which have been well cleansed, and are quite free from any bad smell or taste. The oxygen of the air is previously exhausted by burning a little spirit in the cask, or if its condition is the least doubtful, it is sulphured. Sometimes a small portion of alcohol is now added to each cask, and almost invariably, they also add eight ounces of Catechu, previously dissolved in cold water, to every 100 gallons of Cider. They then fill up, and lightly bung the casks. When the density of the liquor is reduced to 1022, the bungs are to be tightly closed, an "ullage" of one or two inches being allowed to each cask.

THE JERSEY AND CHANNEL ISLANDS method is to let the active fermentation take place in open vessels, covered only by cloth; the scum, or upper lees being removed as it forms. As soon as the liquor becomes clear, and the fermentation subsides, it is casked into sulphured casks, and this process is repeated some three or four times.

When the fruit has been well ripened on the trees, and well mellowed in the heaps, there is generally little difficulty in managing the Fermentation, and still less fear of the liquor not fining properly.

The time over which sensible Fermentation should extend, is necessarily variable, since it depends on the density or richness of the juice, and the temperature of the place. It is most favourable when it is active and regular, but if it is too violent, the liquor will overflow and waste, and if it is too slow, it will be imperfect and develop the disastrous "after fermentation."

THE MANUFACTURE OF PERRY.

"*Perry* is the next liquor in esteem after Cyder, in the ordering of which, let not your Pears be overripe before you grind them; and with some sort of Pears the mixing of a few Crabs in the grinding is of great advantage, making *Perry* equal to the Redstreak Cyder." MORTIMER.

In its earlier stages the making of Perry differs somewhat from that of Cider. The Pears contain more Sugar, and a larger amount of Mucilage. The "must" or rough juice, after pressure, is allowed to remain in open vats, lightly covered, to undergo active Fermentation. As soon as this has subsided, the liquor, between the upper and lower lees, should be sufficiently bright to be drunk off and treated as in the case of Cider; but as a matter of fact, Perry can seldom be made so easily as Cider. The amount of Mucilage renders it necessary, almost invariably, to follow the tedious process of dropping it through bags carefully made of a rather coarse flaxen material, called "forfar." The liquor must be stirred up each time the bags are filled, for the more turbid it is when put into the bags, the brighter it will run through them if the process is carefully managed. The filtered liquor is put forthwith into well-prepared hogsheads. From one to two ounces of isinglass (previously dissolved in some of the cold liquor) is added to each hogshead, the amount varying according to the condition of the liquor and the size of the hogsheads.

The casks are generally placed on their sides, but some think it more safe to place them on their ends, but in either case, an "ullage" to the extent of about a couple of gallons must be left. Then close up tightly and exclude the air, cement the bung, but leave a vent tube through it, the inside end open to the ullage space, and the outside portion bent down and dipped into a cup of water, as before explained. Should the Perry remain quiet for a week, the vent tube may be removed, and the hole it passed through, quickly and effectually closed, or as is sometimes done by the very careful, the tube may be allowed to remain in until spring, though in this case, its outer end must be most scrupulously kept dipped in the water. If the liquor should not remain quiet, and syphon racking into a fresh cask be rendered necessary, it would be a great misfortune for the Perry.

The following abstracts of the methods of Perry making, actually followed in Herefordshire by successful prize winners, have kindly been sent to the Committee:—

Mr. William Hill, Lower Eggleton, Ledbury.—"Pears ripen and rot so quickly, and they heat so rapidly, that they should never be put in a heap. Most varieties should be used as they fall from the tree. With the Barland and other early Pears, it is best to let nearly half fall from the tree and shake the rest down, taking them at once to the mill. Pears require to be very lightly ground, and the New Mill, with stone rollers, is far better than the old one, because it can be set to grind them lightly. The pulp should be placed in an open tub, and allowed to stand a day or two before pressing, then a little skimmed milk, or dissolved isinglass, should be added and well mixed. Allow it still to remain a few days, when it will part in the tub; then filter through bags, and place clear liquor in 100-gallon casks; when three parts full, if the Perry is required for bottling, add a little more isinglass (half an ounce of the best staple, well dissolved in a little bright Perry, for at least four or five days), and whisk well twice a day with a birch rod. Let it stand in the cask 10 or 12 days, if the weather is cool, when it will generally rack off bright and clear, and keep so. If it should not (for Perry making is sometimes very troublesome) it must be racked again into a fresh cask, and more isinglass used, as before. The best time for bottling is April or May.

Barland Perry is sometimes made without dropping through the bags. The liquor from the mill is put into a strong cask, and bunged down. and it generally turns out well."

Mr. John Watkins, Pomona Farm, Withington, Hereford.— " Perry Pears have their season of greatest perfection as well as Dessert Pears. The early varieties should be brought straight from the trees to the mill and ground at once. Some of the later varieties however require to be stored till mellow, or the liquor will be harsh. The pulp should be prepared soon after being ground, and it is best to press all the juice through bags made of forfar shortly afterwards. If this is properly done, it does not require treatment afterwards with isinglass, the same as Cider, but will run bright from the bags. The fermentation is often very difficult to manage, and requires careful watching. If the slightest signs of secondary fermentation takes place, the liquor should be racked into a clean cask, for if allowed to get on the fret, it soon loses its flavour. With these exceptions the treatment required in making Perry is much the same as in making Cider; and the fermentation requires to be regulated on the same principle, and very much in the same manner."

ALCOHOL IN CIDER AND PERRY.—Well fermented Cider of good quality should contain from 5 to 10 gallons of Alcohol to every 100 gallons of the liquor; and the French Chemists say as much as 12 per cent. Good Perry is stated to yield 7 per cent of spirit. The practical rule for estimating the strength of the juice of Apples or Pears, or indeed of all Saccharine unfermented liquors, is to allow 1 per cent of Alcohol for every five degrees of density, as shown by the Saccharometer. For the sake of comparison it may be here added that of the grape vintage, Claret Wine of the first quality should contain from 13 to 17; Sherry, from 15 to 20; and Port Wine from 24 to 26 per cent of Alcohol.

In the early part of last century an extraordinary Cider was made, which received the name of "Royal Cider," and during the wars with France it was extolled to the skies as eclipsing the finest French wines. The whole secret consisted in distilling the Alcohol from one hogshead of Cider and adding it to another; thus making it of double strength, "fortifying" it, as brandy is used to fortify grape wines for exportation.

ORCHARD BRANDY.—A spirit may readily be obtained from the refuse of Apples, or Pears, when it is thought desirable to do so; just as it is from that of grapes after wine making. The cakes from the press are added to the lees in the first racking, with a sufficiency of water, and refermented. As soon as the active fermentation is over, and the lees settled to the bottom, the spirit may at once be distilled from the liquor; or it may, of course, be distilled with better results from the Cider or Perry, after the first fermentation of the must. In either case, the distillation should be effected by means of the water bath, or the brandy will have a burnt rancid taste. The brandy will vary in flavour and strength according to the richness of the must, and the care with which it has been made.

In years of great abundance of fruit, when the barrels are all filled with Cider, and tons upon tons of fruit are still left to rot away in the Orchards, a great economy would be effected if the fruit could be crushed, fermented, and the spirit at once distilled from the liquor; for with good fruit, and ordinary care, a Brandy of good character would be obtained. The great obstacle consists in the uncertainty of the crops. Marshall mentions that "in 1788 there were men who would make 100 hogsheads, that in 1783 did not wet the press;" and it is in the recollection of everybody, that the years 1856-7-8-9 proved a succession of bad seasons, when there was not half a hogshead of Cider made in several of the famous fruit farms in Herefordshire, whereas in 1867-8, after the barrels were all filled, hundreds of bushels of fine fruit were lying in heaps in the Orchards in March. The Apples could not be sold, and were left to decay and be absorbed by mother earth.

THE DIFFICULTIES OF FERMENTATION.

THE DIFFICULTIES OF FERMENTATION.—The combination of circumstances necessary for perfect fermentation, cannot always be commanded by the most skilful managers; but often, it must be added, good fermentation is positively prevented, by sheer carelessness in management. The sources of difficulty are numerous. The season may have been bad, and the fruit not well ripened; the varieties of fruit may be poor, with weak watery juices; the Apples may have been over-heated, or frost bitten, or crushed indiscriminately from the heaps; the prevailing temperature at the time may delay injuriously, or hurry on too quickly, the fermentation; or lastly, there may be a want of cleanliness in the Cider house, the vats, or the implements used. Such circumstances must be expected to result in the production of inferior liquor; but yet with all unavoidable difficulties, good and proper management will prevent the quality of the Cider or Perry from being so bad as it otherwise would have been.

The clear knowledge that Fermentation is due to the growth of certain Fungus Yeast Plants in the fermenting fluid, at once affords the explanation of many of the difficulties that arise in the process, and point out the means best adapted to meet them successfully. Circumstances which encourage the rapid growth of these plants, such as juices rich in Saccharine principle, and a warm temperature, produce a quick active fermentation; whereas their watery juices, deficient in Glucose, cause them to grow so weakly, that a low fretting fermentation sets in, and creates great difficulty, at first to increase its activity, and afterwards to arrest it. Increase of temperature becomes necessary on one hand, and low temperature, and the use of what are called "anti-ferments" on the other. These anti-ferments are now known to stop fermentation by destroying the microscopic plants which cause it. Bearing these facts always in mind, the difficulties most commonly met with, and the remedies they require, will be better understood.

Too Active Fermentation.—When the juice is rich and the weather hot, the fermentation will soon become very active, and may cause both waste and trouble by a copious out-pour from the barrel. In its earlier stages, however, fermentation can scarcely be too active if it is not too long continued ; and all that need be said for the management here, is that everything should be done to cool the temperature ; the windows of the Cider house should be thrown open, wet cloths thrown over the barrels, and water sprinkled about to cause evaporation.

Dilatory Fermentation.—This is a much more frequent and troublesome difficulty, when cold weather sets in suddenly, as it so often does in late autumn ; though it is more often caused by juice of an inferior quality. If, however, it is simply a matter of temperature, and if the tight closing up of the cider house is not sufficient, the introduction of one or two small stoves will be the best remedy. The fermentation may also be aided by drawing two or three gallons of juice from the cask, warming it up to 70° (not higher), and returning it again to the cask, and stirring up the contents freely. The French recommend this stirring up to be done frequently, with a long rod of birch twigs, introduced through the bung hole. There is a fancy sometimes followed of adding a little old Cider or Perry to the cask, and some go so far as to add a little ordinary yeast from malt liquor, but these proceedings are somewhat doubtful and rarely required.

Persistent Fermentation.—The first fermentation will sometimes continue to go on in a subdued form, after its active stage is over. This is called "fretting fermentation." It is the great difficulty to be encountered with juices of inferior quality ; whether this may arise from bad varieties of Apples and Pears, imperfect management of the fruit, or from the indifferent nature of the soil on which the trees have been grown. The French chemists have had much experience in the endeavour to remedy this difficulty, and have obtained an amount of success that demands special notice. They have established the fact that juices of an inferior character are deficient not only in Glucose, but also in Tannin, and Mucilage. When the first fermentation is over, they rack into a cask filled with Sulphuric Acid fumes ; they add half a pound of extract of Catechu (previously dissolved in some of the liquor) to every 100 gallons, which they believe not only assists in fining and preserving it, but also in making it more wholesome; and lastly, they supply the deficiency of Alcoholic Fermentation by the addition of Alcohol in the shape of Brandy to fortify and preserve it.

If the persistent or "fretting" fermentation is allowed to go on, it will exhaust the saccharine principle, and while the liquor loses sweetness and strength, it becomes at the same time more acid. The practical cider maker judges by the smell and taste of the liquor when this period has arrived. The fermentation must now be stopped at once, or the quality of the cider will be still more injured. For this purpose the use of one or other of the anti-ferments—or Yeast Plant destroying agents—must be resorted to, such as Sulphur, Sulphurous Acid Water, Bisulphate of Lime, or Soda, Salicylic Acid, &c. The two first named are the most safe and the most effectual, and indeed they form the base of most of the others used. They are easy of application, economical, and if properly used, ought not eventually to produce any perceptible signs of their presence in the liquor, either to smell or taste. The use of Sulphur, or Salicylic Acid, are the only ones that need be specially alluded to.

SULPHUR.—This agent has been used to arrest fermentation from time immemorial in all the great Wine districts of the Continent, and in all the Cider and Perry districts of England; and indeed it may be said that its use of late years has prevailed universally, wherever the process of Fermentation is carried on. The ordinary mode of its application is very simple. When the liquor is ready for racking, the fresh clean cask is "stummed" or "stunned" as it is termed, (a contraction doubtless of "brimstoned") that is, it is filled with the fumes of burning Sulphur, or Brimstone. A strip of clean canvas cloth, or linen, some ten or twelve inches long by two or three wide, is dipped into melted Sulphur, and then allowed to harden. This cloth match is fixed to a long piece of wire, lighted and passed quickly into the barrel, the wire being fixed by the bung. This soon fills the barrel with the fumes of Sulphuric Acid Gas. The match is removed when it has gone out, from the exhaustion of the atmospheric air, and the fermenting liquor is introduced by Syphon into the barrel, without allowing the Sulphur fumes to escape. The liquor absorbs the Sulphurous Acid Gas, and thus the Yeast Plants are destroyed. It is at first made thick and muddy by the process, but in a short time it becomes clear, and remains so, without retaining the least smell, or taste of Sulphur, if it has been carefully done. Should the fermentation again set in after a few days, as will be known by the hissing noise, the process is repeated. The fumes of Sulphurous Acid Gas are readily absorbed by water, and a saturated solution is sometimes used, instead of the ordinary fumes, from burning the Sulphur in the barrel.

SALICYLIC ACID.—This agent has many advantages as a Yeast Plant destroyer, and it has of late been used more frequently to arrest persistent fermentation. It is a powerful remedy, and requires much care. In proper proportions it is, however, quite harmless, free from smell, or taste. It does not change the colour of any liquor to which it is applied, so long as it is not brought into contact with any metallic substance; but if any iron should be present, and this is the metal most likely to be there, it would give the liquor a black stain. Salicylic Acid can be used in a concentrated solution, and is then more easily applied than Sulphur. An ounce, or an ounce and a half to 100 gallons, is all that is required, and it is simply poured into the liquor immediately after it has been racked. It is an effectual remedy, and leaves no appreciable effects behind it.

Much more might be said on this subject, as for example, about the addition of Bitartrate of Potash, Cream of Tartar, &c., &c., but the attempt to make good liquor from bad juices can never be really successful, and should never be encouraged. The best Cider makers, in the good Cider districts, do not happily require their use, and this axiom may be safely laid down, and deserves to be expressed in capitals, that CIDER AND PERRY IS PURE AND WHOLESOME, IN INVERSE PROPORTION TO THE AMOUNT OF CHEMICALS EMPLOYED IN ITS MANUFACTURE.

A want of Clearness is the last difficulty to be considered, and it is one so very frequent in every quality of fermented liquor, that careful cellarmen seldom trust altogether to Nature, however favourable the process of fermentation may have been. The richer the juice, and the more abundant the Mucilage, the greater is the difficulty of obtaining a clear bright liquor. When the active fermentation is over, and the liquor is racked from the Ices, into a fresh cask, it is customary to add various substances for the purpose of "fining" or clarifying it. To the best qualities of Cider and Perry an ounce or an ounce and a half of Isinglass is added to each hogshead. The Isinglass must be dissolved previously in cold milk, or in some of the cold liquor before adding it to the cask. Fish Glue in about the same proportions will answer equally well. Various other materials are often used, such as powdered charcoal (one pound to the hogshead); the whites of a dozen or two of fresh eggs; roasted apples beaten up; a quart of wheat or barley; and many other heterogenous substances, as chips of Fir, Oak, or Beech wood, a lump of Clay ground up with the fruit in the mill, fresh blood in large quantities, &c., &c., in short, anything that the trade Cider makers can find, which will afford Albumen in a cheap form, and it would not seem to matter much to them, how disgusting the material which contains it may be.

THE PRESERVATION OF CIDER AND PERRY.

"As *Cider* is from time to time a Sluggard, so by like case it may be retained to keep the *Memorials* of many *Consuls;* and these smoaky bottles are the *nappy Wine.*" DR. BEALE, in Evelyn's *Pomona.*

When the liquor is made, and firmly and closely bunged down in the casks, it will improve and keep good for a period, which will vary according to its strength. In former times it was drunk much sooner than it is now. It was never expected to keep long, and would not do so, since very little bottling was practised. The cooling and Summer fruit Cider was ready to drink in a month; that made from the *Gennet Moyle, Pippins,* and *Pearmains* after the first frost ; whilst the *Red Streak* and Winter fruit Cider barrels were not tapped until the winter was well advanced, and were then drank through the following Spring and Summer.

The strongest and best Cider will keep good in casks for four or five years. It was the custom of the last century not to bottle it until two years old, and up to within the last twenty or thirty years, the best Cider was not usually bottled until the late Autumn of the following year, when about a year old. It has now, however, become the general custom, to bottle all Cider and Perry in the early Spring of the next year, and by this means greater richness is obtained, and it comes more quickly into the market; although the risk of loss, from the bursting of the bottles, is greatly increased.

When a cask of Cider or Perry is to be bottled, the bung should be taken out the evening before, that the free gas it contains may escape; and the bottles also should all be filled, if the convenience is present, before any of them are corked, and then the risk of loss from bursting is lessened. The bottles and the corks must be of the best quality, and carefully wired. It is better also, when the cellar space admits of it, to let the bottles remain in sand for a few weeks, or even until the following Autumn, before laying them down in the bins.

A certain amount of insensible fermentation, or molecular change, continues to go on in Cider and Perry, long after all signs of active fermentation have gone by. Thus they improve up to a certain time in the cask, and they will improve still more, and of course last for a much longer time, in bottles. The Alcohol slowly increases at the expense of the Sugar, whilst at the same time the liquor becomes more clear, and acquires additional aroma with its strength. Our ancestors well understood this, as Phillips shows:

> " Cyders in Metal frail improve, the *Moyle*
> And tasteful *Pippin*, in a moons short Year
> Acquire compleat Perfection : Now they smoke
> Transparent, sparkling in each drop, Delight
> Of curious Palate, by fair Virgin crav'd,

> But harsher Fluids different length of time
> Expect : thy Flask will slowly mitigate
> The *Elliots* roughness, *Stirom* firmest Fruit,
> Embottled (long as *Priameian Troy*
> Withstood the *Greeks*) endures e'er justly mild.
> Softened by age it Youthful Vigour gains.
> Fallacious Drink ! Ye honest men beware
> Nor trust its Smoothness ; the third circling Glass
> Suffices Virtue."
>
> <div align="right">PHILLIPS, <i>Cyder</i>.</div>

Nor does the poet in any way exaggerate, either the durability, or the strength of Cider. A supply of good *Foxwhelp* Cider, made in a good year, would have refreshed the warriors for twice, or thrice, or even four times the duration of the siege of Troy. It will retain its full flavour for twenty or thirty years, and a strength moreover, that would require the three permitted glasses to be of moderate size.

Cider or Perry in cask of ordinary quality does not travel well. It is apt to undergo renewed fermentation, and lose all its pleasant qualities. The Cider made in Normandy is much used for seafaring purposes, and the French chemists have had great difficulty to enable it to bear the rolling of the ships at sea. It is with this view in great measure, that, as we have shown, they add Tannin and a small portion of Alcohol to the liquor after the first racking. Economy prevents the addition of sufficient Alcohol to preserve it, and after a number of elaborate experiments, M. Pasteur has proved that the best plan for preserving it safely, is to bring the Cider up to high temperature by artificial heat, and they have established furnaces for this purpose in all their great manufactories. The process however does not improve the quality of the liquor, though it does not render it less effective in preventing scurvy—that dread scourge of seafarers—Good well-made Cider should however travel in cask anywhere in season, and it will safely do so, if its quality is what it generally might be in Herefordshire. In bottles it travels very well in cool weather.

V. THE ORCHARD IN ITS COMMERCIAL ASPECT.

The quantity and value of the Apples and Pears grown in this country are very insufficiently appreciated. The only source from which such information can be obtained, is from the AGRICULTURAL RETURNS published by Parliament, which show that the amount of orcharding in England, that is, "The acreage of arable, or grass land, but used for fruit trees of any kind," was

In 1877 ... ... 159,095 acres.
„ 1880 ... ... 175,200 „
„ 1883 ... ... 185,782 „

The following counties stand highest in the list:

	1877	1880	1883
Herefordshire	24,885	26,683	27,081
Devon	24,776	25,758	26,348
Somerset	20,921	22,993	23,407
Kent	13,097	14,685	17,417
Worcester	14,621	15,854	16,804
Gloucester	11,965	14,178	14,926

Then with a wide difference—

Cornwall	4,497	4,678	4,869
Dorset	3,814	3,716	4,073
Monmouth	2,932	3,618	3,919
Salop	2,944	3,248	3,718
Middlesex	3,051	3,249	3,467

The remainder is divided between thirty-nine other counties.

In Herefordshire, and chiefly also in Devonshire and Somersetshire, the hardy fruits grown are almost confined to Apples and Pears. The increase in the fruit tree acreage for these counties is steady, though not so great; but there is ample room to improve the Orchards that already exist, by supplying the place of the worthless varieties of fruit by those of value.

These Returns afford a sort of basis for calculation, from which a rough estimate may be derived of the value of the Fruit

Crop; but since all the hardy Orchard Fruits are embraced, it will be better to limit the enquiry to Herefordshire, where the fruit acreage is the highest, and where the only Orchard Fruits grown are Apples and Pears.

Herefordshire contains according to the latest Returns 27,081 acres of Orcharding. Of this amount, in these days of cheap and rapid transit, when all apples with size and colour meet with a ready sale as " Pot Fruit," as it is called, that is, fruit for edible or culinary purposes, not less than one sixth must be allowed in this way :— take the product in an average year of 4,514 acres of " Pot Fruit " at the low estimate of 60 bushels to the acre, and at the equally low price of 3s. per bushel, and the value would be £35,626. The remaining five sixths, or 22,567 for the production of Cider and Perry would yield on a very low average two Hogsheads of 100 gallons each per acre; and this at the low price of 3d. a gallon would give £564 17s. 10d., and thus at this computation purposely made so low, the yield from fruit for this County would be at the rate of £3 per acre of Orcharding annually, and if the best fruit was grown and the best Cider and Perry made, as a matter of course, the profit would be much greater.

It must also be remembered that "Pot Fruit" is grown in almost every garden throughout the County, which is not included in the Government Returns. Its amount could scarcely be estimated at less than the Orchard "Pot Fruit," and so an additional sum of £35,626 should be added to the fruit yield of the County, although for the most part it is consumed at home.

The total annual value of the Herefordshire Apple and Pear Crop reaches, according to these estimates, the very large sum of £127,669. As a matter of fact, however, it is not easy to determine the actual produce of English Orchards; for there are no published records of the exact crops they yield year by year. As a general rule the trees of " Pot Fruit," or "Table Fruit " as it is better called, bear a full crop every alternate year, but this is not the case, to the same extent, with the varieties grown for making Cider and Perry. These trees will bear profusely for some two or three years in succession, but after these great " hits " they seem to become exhausted, and, with the exception of a few trees, are apt to yield only a sprinkling of fruit for the next two or three years. This irregular mode of bearing leads to the direct inference, that with proper care, and a good supply of manure, the trees would bear with much greater regularity.

The French have published a few systematic observations on this point. In the Report of the French Congress "*Le Cidre*" is often quoted; it is stated (p.p. 339-40), that M. Varin-Simon, the proprietor of a celebrated Orchard for Cider Fruit, at Yvetot, kept an exact Register of the annual yield from 105 apple trees, for 38 years in succession. His books show that each tree from 5 to 20 years old, gave an annual average over this series of years, of 216 litres (or 40 gallons); and each tree from 20 to 80 years old, yielded 307 litres (or 57 gallons); or taking all the 105 trees during the 30 years preceding 1869, each one gave the annual average of 2 hectolitres, 6 litres (or 45 gallons). This return of course denotes the highest cultivation, good soil, and an excellent climate; but it is still so extremely favourable on the annual average, that we may well believe the popular saying in Normandy "*Le dessus vaut mieux que le dessous*," the trees are more profitable, than the ground beneath them. The actual return, at this rate, would amount to about 10 hogsheads per acre, even if the trees were 60 feet apart, which is double the distance of a thickly planted orchard.

Little information is handed down from early times, as to the Commercial Value of Cider and Perry. Evelyn speaks of *Redstreak* Cider which sold for sixpence the wine quart, "not for the scarcity but for the excellency of it," and he mentions also, that it was sometimes exchanged, on equal terms, for the best French Wines.

In the Household Accounts at Holme Lacy in 1662, the price of the hogshead of cider is set down at £1 14s. 0d., whilst beer cost only £1 4s. 0d. the hogshead.

In a letter dated "Bristol, 20 November, 1691," addressed by one Thomas Wattmore, a Vintner to Sir Barnabas Scudamore " at his seate neare City of Herriford," the writer states, that he bought "six hogshatts of *Red Strike* Sider, and never tasted them at all, but gave you a noate under my hand to pay £25 15s. 0d. for them." The cider turned out badly, and he demands a repayment. At the end of the letter, the writer adds, "I bought 50 hogshatts last yeare at Dimmock and they are as rich as new Canary. I cannot sell bad Sider," &c., &c. This letter gives the price of the famous *Redstreak* Cider in the height of its renown.

The Household Accounts of the Right Hon. James, 3rd Lord Viscount Scudamore, also at Holme Lacy, show that in the years 1703 and 1704 apples were bought at 2s. 3d. the bushel, and in a bill of the time, but without date, a hogshead of *Red Streak* Cider was bought for 10s. 0d.; hogsheads of cider were bought from Amberley and Marden for £1 2s. 6d. each; a hogshead of *Golden Pippin* Cider from Rotherwas cost £1 5s. 0d. It may be mentioned also, that the price of labour for cooperage, cider-making, grafting, &c., set down in these accounts is 1s. per day.

Batty Langley, who wrote at the beginning of the 18th Century (1713), mentions that the *Devonshire Royal Wilding* Cider (a variety that seems to have been lost at the present time), "would fetch five guineas per hogshead, while common cider goeth for 20s." Mr. Hugh Stafford, of Pynes (1753), "has known five guineas refused for a hogshead of cyder from this apple, whilst common cyder sells for 20s., and South Hams from 20s. to 30s."

In Herefordshire, celebrated varieties seem always to have commanded a market, when inferior ones failed to do so. Marshall (1796) mentions *Hagloe Crab* and *Stire* Cider as worth at the press from £5 to £15 the hogshead, but he adds that the ordinary price of Cider "on a par of years" is 25s. per hogshead. In 1720 bottled cider fetched 6d. a bottle, a sum equivalent at the present time to about 3s. In 1825 Mr. John Bosley, of Holmer, (the Mr. John Bosley of the period), sold 12 hogsheads of cider, to be delivered in London, at the price of £12 12s. the hogshead. The cider was made from the *Redstreak*, *Cowarne Red* and *Royal Wilding* apples, which were grown on Holmer Bank, within a mile and a half of the City of Hereford. The twelfth hogshead he had to buy from the then Mr. Davies, of Venn's Green, Marden.

THE ORCHARD IN ITS COMMERCIAL ASPECT.

In Smith's "Dictionary of Commerce" it is stated, that in 1833-4-5 the best cider ranged from 1s. to 1s. 6d. a gallon; family cider for the farmer's own use, or for public houses, 4d. to 10d. a gallon; whilst the cider-kin, or water cider of the labourer when sold ranged from 2½d. to 6d. a gallon; and these prices seem to have amply remunerated the producer.

The market prices of Cider at the present time (1885) are as follows :—For the best quality of Cider sold in cask, from 1s. to 2s. the gallon; and the same quality, when fresh bottled, meets with a ready sale at 8s., 10s., or 12s. the dozen. Cider of the second quality, to which more or less water has been added, sells for family use on draught, from the cask, at from 6d. to 10d. the gallon; whilst the common Cider for farmhouse use will usually fetch £1 the hogshead of 100 gallons. The price of Perry ranges from 4d. to 1s. 6d. the gallon according to quality.

These prices are those which generally prevail immediately after production; but for the Cider made from special varieties of fruit, and for the best Cider a few years in bottle, the prices are much higher. At a public auction, a short time since (1880), at the late Mr. Mason's, *Foxwhelp* Cider was sold freely at 30s. the dozen, and *Taynton Squash* Perry fetched 28s. a dozen, at the same sale. Either of these varieties, and some others too, when of good age and of the first quality, will always command high prices. The *Foxwhelp* Cider from Mr. John Bosley, of Lyde, near Hereford, which won the First Prize at the Herefordshire Agricultural Society's Meeting at Ledbury, in 1884, sold quickly at £1 the dozen. Oldfield Perry, in a good season, has been sold for a guinea a dozen, from the glebe land in the parish of Credenhill.

As a general rule the small orchardists make better Cider than the large farmers, and for the very reason that they give their chief thought to it. It is their main harvest, and it is not too much to say, that many of them get from their trees not only the rent they pay, but, in addition, a considerable help towards their livelihood. A rough calculation may easily be made for an acre of orcharding well cared for, and fit, by the pigs and fowls constantly beneath the trees:—at 30 feet apart there will be 50 trees to the acre, and with a fair "hit" of fruit, 40 of them should yield, at the very least, six hogsheads of liquor. This Cider or Perry at the rate of 6d. a gallon will bring £2 10s. the hogshead, or £15 altogether; but some of it should be worth more than this. Then there is the "Table Fruit" from the remaining 10 trees still left, to be sold in the market through the Autumn and Winter; and, in addition, the profit that may be derived from the produce of the ground. An acre or two at this rate would give a handsome return; and if the occupier be a smith, a tailor, a shoemaker, or a blacksmith, as not uncommonly happens, this addition to his trade earnings will put him in easy competence, and enable him to educate and place out his family to advantage. There is one other important "if," however, and that is, if he does not drink too freely from his own vats.

The fruit trees, on farms of higher pretentions, are sometimes made important sources of income, and should certainly contribute much more towards the rent, than they usually do.

VI.—THE RENOVATION OF THE ORCHARDS.

The condition of the orchards generally, at the present time, is most unsatisfactory, and close attention will be required for many years, to restore their value. A Century of neglect, has caused the loss of many of the best varieties of fruit, for the number of vacancies, from the prevalence of cold wet weather, the ravages of insects, the violence of storms, or the effect of age, that are constantly occurring in the Orchards, is very great. These vacancies must be filled up, by the conditions of the occupier's lease, and the young trees for this purpose, seem to have been procured hap hazard, that is, at the least possible expense and trouble, and thus a large number of chance seedlings, unproved and worthless varieties, have found their way into the Orchards. They are without names, and for the most part do not deserve a name.

The first step towards the improvement of the Orchards will be, to subject them to a gradual and thorough revision. Stock should be taken of every individual Apple and Pear tree on the farm, and its character and condition carefully considered. Such trees as are mere cumberers of the ground should be cleared off at once, root and branch; and such varieties as are proved to be unmistakeably inferior, should have their places supplied by those which are known to be good. If the trees of inferior kinds are vigorous and healthy, they should be cut back and grafted on all the branches. Every spur of not more than two inches in diameter, should be grafted with strong growing scions, so as to bring them into bearing again quickly, with the loss of only two or three seasons; but if the condemned trees are old, they should be up-rooted. Every renewed tree, whether by grafting or planting, should be of a well proved variety, since it must never be forgotten that, WHEN ONCE PLANTED, THE BEST FRUIT TREES DO NOT REQUIRE MORE CARE OR EXPENSE THAN THE WORTHLESS ONES.

A complete revision of the Orchards will require some years to effect, but it is a work of great interest, and will well repay by success, all the time given to it.

TEST OF THE QUALITY OF THE TREES.—The commercial value of any fruit for the manufacture of Cider and Perry, depends almost entirely on the density, or richness of its juice. This most important condition may be definitely ascertained, with a little experience, by anyone who will take the trouble to do so. It is only necessary, in a good season, to crush out the juice from five or six well-ripened apples of the variety it is desired to test; filter it through white porous paper; and having procured a small instrument called a Saccharometer, to float it in the fresh juice. The scale marked on the instrument will give the density of the juice, as compared with the standard of distilled water placed at 1000. This density is chiefly caused by the saccharine matter the juice contains, and chemists have ascertained, that in general terms, every five degrees of density shown on the scale of the Saccharometer, denotes a spirit-producing power equal to one per cent., or in other words, one gallon of spirit in one hundred of the liquor examined. Now, moderately good Cider should not contain less than six per cent. of alcohol ; and about one-fourth or one-sixth of its density should still be left of unreduced sugar, to give it sweetness and body. The density of juice, therefore, for moderately good Cider should not be less than 1040. The richer the juice, the higher the density, and the greater its value. Juice which has a density below 1040 though it may make Cider, or Perry, if it has been grown on good lands, can never give the superior quality, it is so desirable to produce. THE SACCHAROMETER WILL THUS POINT OUT ALL THE VARIETIES OF FRUIT TREES WHICH SHOULD BE UPROOTED. The instrument requires a little experience to use it rightly, but is yet very simple.

The following table shows, at a glance, the exact amount of spirit-producing power contained in juice of any given density, according to the experiments of the French Chemists :—

Table shewing the amount of Sugar contained in the French Litre of fresh Apple Juice, and the per-centage of absolute Alcohol it will produce on Fermentation. (The Litre is equal to 1 ¾ pint English, or 35 oz.) Baune's Densimoter.]   [Extracted from *Le Cidre*, p. 130.

Density of Juice.	Sugar in 35 oz.	Alcohol per Cent.	Density of Juice.	Sugar in 35 oz.	Alcohol per Cent.	Density of Juice.	Sugar in 35 oz.	Alcohol per Cent.
1010	1.098	1.138	1038	2.273		1066	5.124	
—12	1.141		1040	2.350	4.85	—68	5.181	
—14	1.183		—42	3.006		1070	5.274	9.50
1015	1.203	1.93	—44	3.084		—72	5.372	
—16	1.223		1045	3.114	5.64	—74	6.013	
—18	1.247		—46	3.161		1075	6.051	10.51
1020	1.322	2.48	—48	2.252		—76	6.090	
—22	1.353		1050	3.332	6.43	—78	6.183	
—24	1.400		—52	3.409		1080	6.261	11.33
1025	1.415	3.3	—54	4.065		—82	6.338	
—26	1.431		1055	4.102	7.26	—84	6.431	
—28	2.040		—56	4.143		1085	7.030	12.14
1030	2.071	3.57	—58	4.220		—86	7.072	
—32	2.118		1060	4.308	8.11	—88	7.149	
—34	2.114		—62	4.386		1090	7.242	12.98
1035	2.169	4.12	—64	5.031		—92	7.324	
—36	2.195		1065	5.077	8.76	—95	8.210	13.86

It will be observed that the sugar increases more in relative amount, the higher the density becomes; which is explained by the fact, that the Mucilage, to which the density in the lower ranges of the scale is partly due, does not increase as the sugar does, in the higher ranges of density. The alcohol in this scale is a little more than 1 per cent. for every 10 degrees of density, instead of 1 per cent. for every 5 degrees, which is the scale of the English Excise Offices.

TABLE FRUIT.—The varieties of fruit suitable for cooking or dessert purposes—"Pot Fruit" or "Table Fruit," should be grown much more frequently than is usually the case, where the Orchards are near home and can thus be protected. Size and colour are essential for market purposes, and the longer the fruit will keep, the more valuable it will become. As much as £10 an acre is not unfrequently given for fruit of this character, in the homestead Orchard, but the market value must of course depend very much on the season; as a general rule "Table Fruit" is not well adapted for making Cider. The French have the proverb "*Petites pommes gros Cidre*," "small apples, rich cider," and so too the finest Wines are produced from the smallest grapes. Large apples have too much Mucilage by themselves, though when the markets are overstocked, they are not unfrequently added with advantage to the smaller ones for making Cider.

THE CIDER HOUSE.—The want of suitable buildings is a very serious drawback to the proper storage of fruit, and to the manufacture of Cider and Perry in perfection. Marshall and other writers have pointed out the saving in time and labour that would be effected, if every Orchard Farm had a well-arranged Fruit and Cider House, furnished with simple machinery, and with suitable mechanical fittings. Such buildings should be so constructed as to command a low, or, still better, different degrees of temperature at will. They need not necessarily be expensive. Thick walls of stone, or hollow bricks, and a good thick straw thatch, with due arrangement for free ventilation, is all that is essentially required. With these advantages, it would be quite possible to regulate and prevent those sudden changes of temperature which so frequently prevail in Autumn, and which are often so injurious to the liquor; at one time suddenly checking fermentation, and at another exciting it again, when it should be cool and quiet.

In America, great advantage is derived from the refrigerating houses, used by the fruit growers; by means of simple and ingenious mechanical contrivances, they preserve their Apples and Pears, at a temperature a little above freezing point, in the finest condition, so that they are ready for the market at any time. In the manufacture of Cider and Perry, these houses also afford the utmost advantage. The details of their construction and management, are given in full in Downing's *"American Orchardist"*; and when it is considered, that these appliances are only required during the late Autumn and early Winter months, it should be a matter of serious consideration for the landlord and tenant, whether the advantage of such buildings should not be provided for a Fruit Farm.

Mr. Thomas Andrew Knight, at the commencement of the present century, felt so much the necessity of commanding a low temperature for his Cider, that he built a cellar, on the hill side at Wormesley Grange, in the bed of a small stream, so that he could at pleasure keep it filled with running water, and thus check any tendency to second fermentation. The theory was good, but the practical inconveniences connected with this means of carrying it out, proved to be greater than the advantages derived from it.

DISTRICT FACTORIES.—The establishment of large Cider and Perry Factories, in the immediate vicinity of the Orchards, has been often advised. Marshall and other old writers recommended them, and it is very probable that they would have been established much more generally, if the causes which produced such a lamentable neglect of the Orchards had not prevented it. There are private Cider and Perry makers now, who will buy up the superior varieties of Apples or Pears they require, but they will not purchase at any price, the enormous amount of poor fruit, which at present pervades the Orchards. The farmers, therefore, have to make the Cider and Perry themselves, as best they can, and sell it in bulk, at a very low price, to the ordinary "Cider

Merchants." From their hands it passes on, if it will bear the saccharometer test, to other manipulators, and, eventually, it is believed to reappear as Hock, Champagne, Sherry, or Port, as may be required in commerce at the time.

The establishment of Cider and Perry Factories, would prove of the greatest advantage in the Orchard districts. A ready home market for the best kinds of fruit, would lead to the gradual extinction of the inferior varieties, and the manufacture of Cider and Perry of superior quality, would soon cause these wholesome beverages to be properly appreciated, and the outer world to value their high character. Under present circumstances, when a great "hit" of fruit occurs, the Apples and Pears are scarcely saleable at any price, the home barrels are all filled, and the waste is enormous. It sometimes happens at these times, that a barrel of Cider is placed in the yard ready tapped, with a mug at hand, that all comers to the house may help themselves. Such prodigal hospitality is by no means desirable, and if the demand for good Cider was as great as it might be made, its value would soon put a stop to such wasteful use.

It is precisely in these good seasons, when fruit is so abundant and well ripened, that the best liquor can be made. It would be the golden opportunity for a Factory, supported by capital. Very large quantities of Cider and Perry could be made, and laid by in cask and in bottle, to meet the failure of succeeding years. With good management, a company formed for the manufacture of Cider and Perry, could scarcely fail to give a very handsome return to the proprietors, and at the same time, it would greatly increase the value of the Orchards.

VII.—ORCHARD PROSPECTS.

English agriculturists have now to meet the competition of the world, and it is desirable on every account, that they should enlarge their sphere of action. Instead of confining themselves so much to Corn and Cattle, as they have hitherto done, they should pay closer attention to the growth of other products, which will command a constant and lucrative market, in our own populous and wealthy towns; such as Hops, where the soil is suitable; Poultry and Eggs; Milk, Butter, and Cheese; Fruit of all kinds; and such Vegetables as local circumstances may require, or good judgment determine. Happy in these times are they, who, living in districts especially adapted to the growth of hardy fruits, can turn their efforts in this direction. Our Orchards ought to supply economically and profitably, the markets of our cities and towns with an abundance of Apples and Pears, and to be able to meet successfully, moreover, an active competition from the Continent of Europe, from America, and even from Australia. It is true that the rent of land is dearer, and the fruit seasons much more uncertain in England; but these disadvantages are almost balanced by the greater expense of labour—at least in America—our greatest rival of late years; by the additional expense of packing; the cost of carriage; the liability to injury; and by the still more serious item of profit to the middlemen or importers. The importation of fruit must always be more difficult than that of grain, and the cost greater; this cost, moreover, must increase as soon as the commercial depression of the last few years passes away, and ship freightage returns to its ordinary rates. There is every reason, therefore, to believe that steady perseverance in Orchard culture will meet with a successful reward.

The occurrence of favourable seasons, affords the greatest opportunity for remunerative Orchard management. At these times, in addition to increased cellar storage for vintage fruit, and the sale of fruit in the market, there is great scope for individual energy in the preservation of table fruit. This may be done in a variety of ways. Apples and Pears may simply be dried whole; as, for example, the *Herefordshire Beefing*, the *Norfolk Beefing*, &c., &c.: they may be pealed, covered, and the flesh dried in the shape of "Apple Chips," "Apple Cuttings," or "Apple Rings," as the Americans call them. They may be preserved in syrup in tins, or better still, they may be converted into jelly. All these modes of preparing fruit for sale, do not require any great capital; and if the preparations are well made, they give a good profit, and keep well, to supply the deficiencies of the first half of the year, when fruit is scarce.

Herefordshire, Devonshire, and Somersetshire, and other districts capable of producing Cider and Perry of good quality, have a peculiar advantage, in the possession of a branch of agricultural industry, that may be made very remunerative. It is one the least likely to be interfered with, by the fluctuations of ordinary trade, and has therefore with proper care, only the seasons to contend with. The present state of our legislature is most favourable to its extension, since there are no longer any restrictions on its produce by taxation ; nor yet on its sale direct from the Orchards ; whilst as regards foreign competition, there is no probability that the supply for our home consumption can be seriously interfered with, for this, if for no other reason, that beverages which only contain so slight a proportion of alcohol, are readily susceptible of re-fermentation, caused by the constant shaking, incident on conveyance from a distance.

HOME FRUIT MARKETS.—The authorities in the city of Rouen, in the year 1884, established a fruit market. It would greatly conduce to the improvement of Orchard culture, if the Agricultural Societies in the special fruit districts, would take up the subject, hold annual Exhibitions of fruit, and offer a schedule of prizes. Agricultural meetings are almost always held conveniently in the Autumn, and such exhibitions of fruit could scarcely fail to prove attractive, and they would certainly spread a knowledge, which would lead to the growth of the superior varieties of fruit.

The theory and practice of Horticulture and Fruit Growing, might also be introduced judiciously, with great advantage, as a science subject, in our rural Elementary Schools, as was most successfully done some years since by the late Professor Henslow, in his village school in Cambridgeshire. In these respects, English schools are far behind those on the Continent. There, elementary instruction in Horticulture, is aided as it should be, by manual work in the garden ; and to instruction in the growth of vegetables, herbs, and flowers, is added the practice of budding, grafting, and pruning fruit trees. This excellent practice could not fail to produce a much more extended interest, in the production of the best varieties of fruit ; as the knowledge of how to bud a rose briar, has introduced many of the most beautiful roses into Herefordshire cottage gardens.

Landlord and tenant are alike interested in the utmost development of our home industries. The greatest attention must be paid to the special products of every district. Great competition must be met by high cultivation, by economy, and by intelligent persevering industry. The land must be managed, if not in the letter, yet in the economic spirit of John Stuart Mill, who pointed, as an illustration, to the cabbage of the French proprietor, so carefully dug round, watered, and manured ; so individualised, in short, as though the whole profit of the farm centred in that one single vegetable. By thus paying greater attention to minute details, the farm may become, what it ought to be, in these days of competitive agriculture, in both hemispheres—a duplicate of the garden on a large scale.

A flask of prime Cider is the crowning enjoyment, in Tennyson's charming description in " *The Pic-nic* " :—

> " There on a slope of orchard, Francis laid
> A damask napkin, wrought with horse and hound ;
> Brought out a dusky loaf that smelt of home,
> And, half cut down, a pasty costly made,
> Where quail and pigeon, lark and leveret lay,
> Like fossils of the rock, with golden yolks
> Imbedded and injellied ; last with these
> A flask of Cider from his father's vats
> Prime, which I knew ; and so we sat and ate."

HENRY G. BULL, M.D.
CHARLES HENRY BULMER, M.A.
J. GRIFFITH MORRIS.

REPORT OF THE COMMITTEE APPOINTED TO ATTEND THE CONGRESS OF THE POMOLOGICAL SOCIETIES OF FRANCE, ON BEHALF OF THE WOOLHOPE CLUB, HELD AT ROUEN, FROM OCTOBER 2ND TO THE 12TH, 1884.

Your Committee, having obtained the Schedules of the Exhibition to be held at Rouen, thought it best to compete in the classes open to strangers. A collection of Table Fruit was therefore obtained from the gardens of Stoke Edith, Holme Lacy, Thing-hill, and other places. It consisted of fifty-seven varieties of Dessert Apples, fifty-seven varieties of Culinary Apples, and thirty-six varieties of Pears. This collection was very fine. It formed the leading attraction at the Exhibition, in the Hall of the Hôtel de Sociétés Savantes at Rouen, and a Gold Medal was awarded to it, by the SOCIÉTÉ CENTRALE D' HORTICULTURE DE LA SEINE INFÉRIEURE.

A fine bunch of Black Alicante Grapes from the gardens at Eastnor Castle was also taken, and received from the same Society a large Silver Medal.

The collection of Orchard Fruits exhibited by the Woolhope Club, consisted of fifty-six varieties of Cider Apples, and forty-two varieties of Perry Pears. To this collection the ASSOCIATION POMOLOGIQUE DE L'OUEST awarded a Bronze Medal.

Two varieties of Cider made from mixed fruits, and four varieties of Cider made from a single variety of fruit, with two varieties of Perry, were also exhibited. To the Cider from mixed fruits, a Silver Gilt Medal was given, and to that from a single fruit, a Silver Medal. Prizes were not offered for Perry, of which very little is made in Normandy.

The first six Parts of THE HEREFORDSHIRE POMONA, were also exhibited, and the high distinction of a "Diplôme d'Honneur" was awarded to the Woolhope Club, from the SOCIÉTÉ CENTRALE D'HORTICULTURE DE LA SEINE INFÉRIEURE, for the Table Fruits represented in the Work; and a second was also given by the ASSOCIATION POMOLOGIQUE DE L'OUEST for the Vintage Fruits.

A Gold Medal was also specially awarded to Dr. Hogg, for his life-long work in Pomology.

The receipt of these high honours, did not cause your Committee to forget, that the chief objects of their visit to Rouen were, first, to ascertain whether the Apples called "Norman" in Herefordshire, were really Norman varieties; and secondly, if they were not so, to select a few of the most valuable varieties from the Norman Orchards, to introduce into Herefordshire.

Eighteen of the best so-called Norman Apples of Herefordshire, were placed together on the exhibition tables at Rouen. Your Committee carefully compared them with the three thousand plates of Vintage Fruits present: the attention also of the leading exhibitors from Normandy and Brittany, was specially called to them; but, with one exception, they were quite different to all others there, and were unknown to the Norman nurserymen and growers. The exception was the "*Foley Norman*," which local tradition states to have been introduced into Herefordshire, by Mr. Edward Thomas Foley, of Stoke Edith (c. 1810-20). This Apple was the same as the *Blanc Doux* of the Rouen Catalogue, but it is one that has not borne well the modern test of exact analysis, and it has therefore lost much of its repute in the Norman Orchards.

Your Committee next proceeded to select a few of the best real Norman varieties, to be introduced into Herefordshire. They decided that the following characteristics were essentially necessary.

1.—The fruit must possess the very best quality of juice.

2.—The trees must be hardy, vigorous, and fertile.

3.—They must blossom at varying intervals.

4.—The fruit must attain maturity in late autumn or winter.

And 5.—They must have obtained the highest repute in the Norman Orchards.

With the kind assistance of Monsieur A. Hauchecorne (one of

the distinguished authors of the great French work "*Le Cidre*"); Monsieur Michelin, of Paris (one of the original promoters of the Congress appointed by the French Government for the study of Cider Fruits); Monsieur Héron (President of the SOCIÉTÉ CENTRALE D'HORTICULTURE DE LA SEINE INFÉRIEURE); Monsieur Legrand, Nurseryman at Yvetot; Monsieur Lesueur, of Rouen, and other Norman growers of Cider Fruits; your Committee have selected eight varieties, which meet all the requirements laid down for their guidance.

The Apples selected were ARGILE GRISE, BÉDAN-DES-PARTS, BRAMTOT, DE BOUTTEVILLE, FRÉQUIN AUDIÈVRE, MÉDAILLE D'OR, MICHELIN, and ROUGE BRUYÈRE.

Trees of all these valuable varieties of true Norman Apples have been sent to Messrs. Cranston and Co., King's Acre, Hereford, who will propagate carefully from them. It is believed that they will prove very valuable in the orchards of Herefordshire.

The table on page 90 gives a summary of their virtues.

Your Representatives, in conclusion, desire to express their sense of the great kindness and courtesy shown to them during their visit to Rouen.

ROBERT HOGG,
GEO. H. PIPER,
HENRY G. BULL.

October, 1884.

NORMAN CIDER APPLES, INTRODUCED BY THE WOOLHOPE CLUB, 1884.

HABIT—CHARACTER—ANALYSIS OF JUICE, &c., FROM THE CATALOGUE OF THE SCIÉTÉ CENTRALE D'HORTICULTURE DE LA SEINE-INFÉRIEURE.

No. on Plate	Name.	Time of Blossoming.	Ripe.	Flavour of Fruit.	Quality of the Juice.	Density of Juice. (Sp. Gr.)	Useful Properties contained in 1 Kilogramm (35 oz., 100 gr.,) of Juice.				Character of the Tree.
							Sugar Grammes (15½ gr.)	Alcohol per 100.	Tannin Grammes (15½ gr.)	Acidity Grammes (15½ gr.)	
1	Rouge Bruyère	Beginning of May.	November.	Slightly bitter, excellent.	High in colour, very good flavour.	1,075 to 1,080	175	10 to 11	7,000	0,001	Healthy and fertile, round headed growth.
2	Braintot	Beginning of May.	November.	Slightly bittersweet, excellent.	Good in colour, perfume, and taste.	1,092 to 1,105	226	13 to 14	6,000	1,070	Healthy, vigorous and fertile, of handsome growth.
3	Médaille d'Or	Beginning of June.	November.	Bitter, excellent.	Good in colour and perfume.	1,102	238	14 to 15	5,509	1,428	Very vigorous and fertile, with upright growth.
4	Bédan-des-Parts	End of April.	December.	Bitter sweet, excellent.	Very good in colour, taste, and perfume.	1,084	197	12	5,000	1,070	Horizontal growth, healthy, vigorous, and fertile.
5	Michelin	End of May.	December.	Sweet and good.	Good in colour, flavour, and perfume.	1,080 to 1,083	190	11 to 12	5,509	1,071	A round headed tree, healthy, vigorous, and fertile.
6	Argile Grise	Beginning of May.	November and December.	Slightly bitter, very good.	Good in colour, perfume, and taste.	1,075 to 1,080	194	10 to 11	5,509	1,071	Upright in growth, vigorous, and fertile.
7	De Boutteville	Beginning of May.	December.	Sweet and good.	High in colour, perfume, and flavour.	1,083	183	11 to 12	6,000	2,142	Upright in growth, healthy, and very fertile.
8	Fréquin Audievre	End of May.	December.	Bitter sweet, excellent.	High in colour, good taste, and perfume.	1,079	180	10,5	5,509	1,320	Horizontal growth, vigorous, and fertile.

AN ALPHABETICAL LIST
OF THE
MOST ESTEEMED VARIETIES OF
CIDER APPLES.

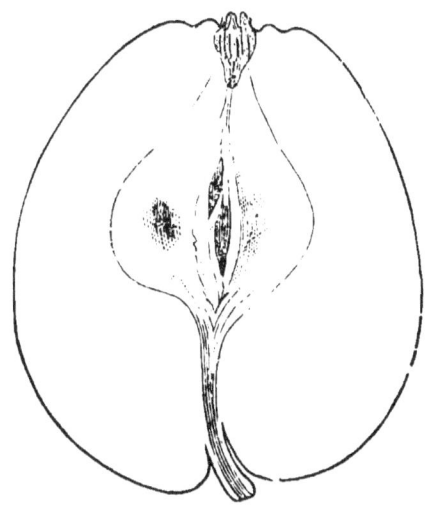

ARGILE GRISE.

The *Argile* is one of the oldest varieties in the Norman orchards. Its origin is unknown, but it has long been highly esteemed in all the Departments of the North-west of France, in which Cider is produced. Its name is so popular, that it has been given to many varieties, and often to those of inferior value. The *Argile Grise* is the best of all the varieties.

Description.—Fruit: rather below the middle size, ovoid, with obtuse angles as it narrows towards the eye; often fuller on one side than the other. Skin: greenish yellow, more or less covered with a thin grey russet; it sometimes takes a pale tinge of red colour on the sunny side. Eye: small and closed, with short, broken sepals, seated in a shallow cavity, with folded margins, and small tubercles between the folds. Stalk: small and short, frequently connected with the fruit by a fleshy prominence on one side. Flesh: yellow and tender. Juice: plentiful, slightly bitter, but still sweet and pleasant.

"The *Argile Grise* belongs to the *Fréquin* group of Cider Fruits," says Monsieur Hauchecorne; "it is equally valued in the orchard with *Rouge Bruyère*," and is believed to make cider of the best quality. The juice has a good colour, and a density of 1·075, and sometimes more. One thousand parts contain of alcoholic sugar 194; tannin, 5·509; mucilage 15; acidity 0·920; salts, &c., 3·571; and water 781.

This variety was introduced from Normandy into Herefordshire, by the Woolhope Club in 1884, and has yet to be tried in Herefordshire.

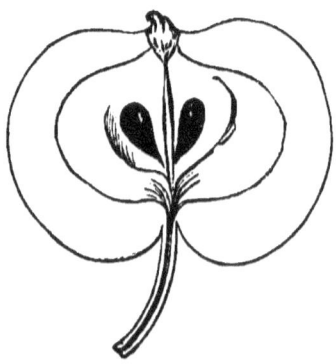

BASTARD FOXWHELP.

There are two or three small apples called by this name, but that which is the most esteemed and grown, is figured here.

Description.—Fruit: small and oblate, sometimes somewhat roundish, even and regularly formed. Skin: smooth and shining as if varnished, entirely covered with bright crimson, and striped with darker crimson on the side exposed to the sun; but on the shaded side, it is greenish yellow striped with crimson; the stalk cavity only is lined with russet. Eye: very small and closed, with short connivent segments, placed in a shallow saucer-like depression;

tube, conical; stamens, marginal. Stalk: very long and slender at its insertion, and throughout its length, but thicker at the end, inserted in a deep cavity. Flesh: yellowish stained with red, firm, unusually acid. Cells of the core, slightly open; cell-walls, orbicular.

The chemical analysis of the juice of the *Bastard Foxwhelp* (season 1876), by Mr. G. H. With, F.R.A.S., F.C.S., Trinity College, Dublin, gave the following results :—

Density of fresh juice	1·042
Ditto after 24 hours' exposure to air	1·042
100 parts of juice by weight, yielded of	
Sugar	7·780
Tannin, Mucilage, Salts, &c.	4·335
Water	87·885

The *Bastard Foxwhelp* bears well, and is much esteemed by some growers, who think they detect in the cider which it helps to make, a slight *Foxwhelp* flavour.

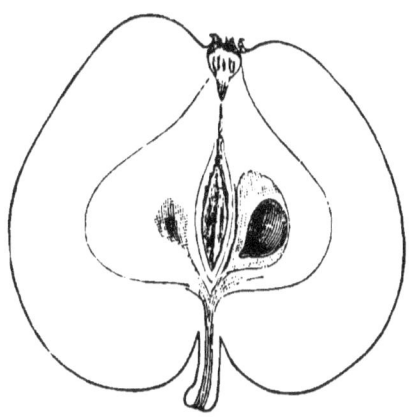

BÉDAN-DES-PARTS.

The *Bédan, Bédengue, Bec d'Âne*, with other varieties in name, has held a high repute in the Norman Orchards, from time immemorial. This particular variety, is superior to all the *Bédans*, in the richness and colour of its juice. It is a seedling grown by Monsieur Legrand, of Yvetot, which first bore fruit in 1874.

Description.—Fruit: small, broad at the base, often larger on one side. Skin: pale yellowish green, with a clear red cheek on the side next the sun; small grey spots are scattered over the surface, and sometimes brown patches. Eye: small and closed, set in a shallow, irregular cavity, with grooves and small tubercles between them. Stalk: strong, half an inch long, inserted in a shallow, narrow cavity, which is lined with russet, which russet extends, more or less, over the base of the apple. Flesh: yellowish, tender, and juicy, slightly bitter in taste, but with good flavour. Juice: highly coloured.

"This new variety," says Monsieur Hauchecorne, "takes a high place among fruits of the first quality, from the fertility of the tree, the high colour of its juice, and its richness in sugar, tannin, and aroma." The density of the juice is 1·084. One thousand parts contain of alcoholic sugar 195; tannin 5; mucilage 10; acidity 1·070; salts, &c., 1·030; and water 776.

This variety was introduced from Normandy into Herefordshire, by the Woolhope Naturalists' Field Club in 1884, and has yet to be tried in our Orchards.

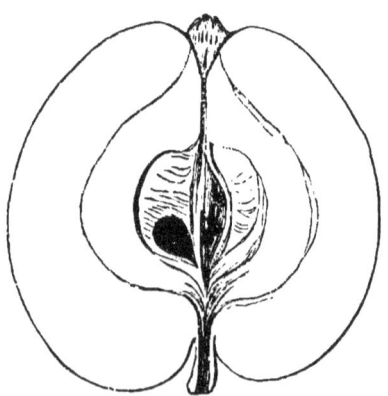

BLACK FOXWHELP.

[Syn: *Monmouthshire Foxwhelp.*]

This apple is very widely grown throughout the county, and is to be found in the majority of "apple heaps." Its definite ovate shape, smooth surface, and dull colour, make it quite unmistakeable. It bears very freely, and this, perhaps, is its best qualification, for the cider made from it, is thin and poor.

Description.—Fruit: small, roundish ovate, inclining to short conical, even in its outline, slightly angular towards the crown, where it is prominently plaited round the eye. Skin: smooth and rather shining, of a dark mahogany colour next the sun, but on the shaded side, it is greenish yellow, covered with broad broken stripes of bright crimson. Eye: small and rather open, with rather connivent segments, and set nearly on a level with the surface, with only a very slight depression; tube, short conical; stamens, medium. Stalk: short, set in a shallow cavity. Flesh: yellowish, sometimes with a greenish tinge, briskly acid. Cells of the core, open; cell-walls, obovate.

The chemical analysis of the juice of the *Black Foxwhelp* (season 1876), by Mr. G. H. With, F.R.A.S., F.C.S., Trinity College, Dublin, gave the following results:—

Density of fresh juice … …	1·038
Ditto after 24 hours' exposure to air	1·048
100 parts of juice by weight, yielded of	
Sugar … … … …	6·400
Tannin, Mucilage, Salts, &c. …	5·206
Water … … … …	88·394

The *Black Foxwhelp*, notwithstanding its small amount of sugar, is still esteemed in some orchards, when mixed with sweeter varieties, for the amount of tannin it contains. The results of the analysis, show it to be a variety of little value.

The tree is hardy, grows upright and bears well.

The sooner the trees of the *Black Foxwhelp* are re-grafted, or cut down, the better.

These several apples bear the *Foxwhelp* name. They have no special history, but the inference is, that they are, what tradition supposes them to be, seedlings from the *Foxwhelp*.

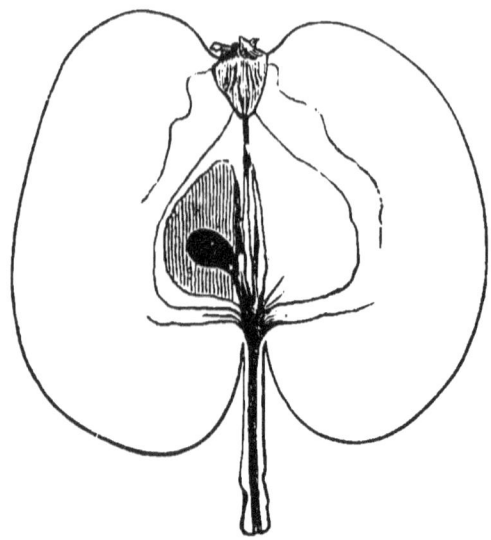

BLACK HEREFORD.

[Syn: *Black Norman.*]

This very distinct variety is again without history, and was not known by any of the Norman exhibitors, at the Rouen Exhibition in 1884.

Description.—Fruit: roundish and flattened, obscurely ribbed, especially round the eye. Skin: smooth and shining, unctuous to the touch, after the fruit has been gathered; dull mahogany red, on the side next the sun, and gradually becoming paler towards the shaded side, which is deep green, and slightly mottled with red. Eye: closed, with long, leafy, convergent segments, set in a rather deep irregular basin; tube, conical; stamens, medium. Stalk: long and slender, inserted in a deep, wide, funnel shaped cavity, which is slightly russety. Flesh: greenish, very tender, juicy and brisk, with a faint sweetness. Cells of the core, quite closed; cell-walls, ovate.

The chemical analysis of the juice of the *Black Hereford* (season 1878), by Mr. G. H. With, F.R.A.S., F.C.S., Trinity College,

Dublin, gave the following results:—

Density of fresh juice	1·036
Ditto after 24 hours' exposure to air	1·037
100 parts of juice by weight, yielded of	
Sugar	11·905
Tannin, Mucilage, Salts, &c.	1·125
Water	86·970

The *Black Hereford* is a favourite in the orchards. It is a late fruit, and is thought to make a strong, rich, long-keeping cider, with a peculiar flavour.

The tree is hardy, blossoms towards the end of May, and ripens its fruit the end of October. It is hardy, and bears well.

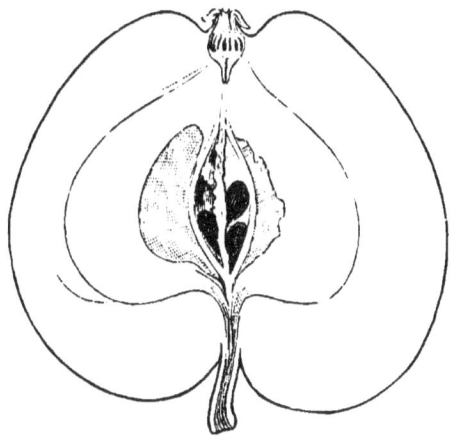

BRAMTOT.

A seedling grown by Monsieur Legrand, of Yvetot, Seine Inférieure. It first fruited in 1856, and was named after Monsieur Bramtot, a manufacturer of Yvetot. It is thought to be a seedling, from the old variety, *Martin Fessard*. Introduced into Herefordshire, by the Woolhope Club, in 1884.

Description.—Fruit: of middle size, symmetrical, but sometimes with unequal sides, wide, and flattened at the base, but contracted towards the eye. Skin: clear yellow, with a touch of carmine towards the sun, its surface being scattered over with numerous grey spots. Eye: small and closed, with long reflected sepals, and placed in a very narrow cavity, with grooved sides. Stalk: short, thin, and woody, set in a narrow, deep cavity. Flesh: whitish yellow, and tender, with an abundant juice of a sweet and pleasant, though slightly bitter flavour.

"This excellent variety," says Monsieur Hauchecorne, "both in tree and fruit, possesses virtues as an apple for the press, which are rarely united in so high a degree." The juice is of good colour, and has a pleasant aroma. Its density is as high as 1·092, and in good seasons it reaches 1·105. A kilogramme contains 226 grammes of sugar, which gives an alcoholic strength from 13 to 14 per cent. There are also 6 grammes of tannin, and 1·070 of acidity, in each kilogramme of juice.

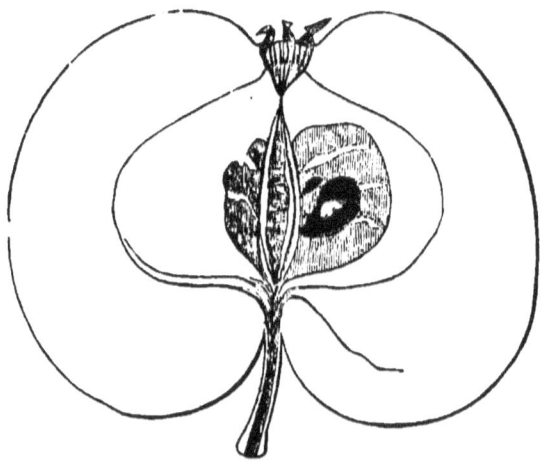

BRAN ROSE.

The origin of this variety is not known. It is a favourite apple in the Herefordshire orchards, and is widely grown throughout the county.

Description.—Fruit: medium size, roundish oblate, with five or more obtuse angles at the side. Skin: almost entirely covered with red, becoming much darker on the side towards the sun, and everywhere shewing numerous small yellow spots Eye: partially open with reflexed segments, set in a narrow basin, more or less irregular. Stalk: slender, more than half an inch long, and set in a deep and narrow cavity, lined with russet. Flesh: deeply stained red, from the skin to the outside core lines. Juice: plentiful, of a deep rose amber colour, sweet, with some roughness of taste.

The chemical analysis of the juice of the *Bran Rose* (season 1883), by Mr. G. H. With, F.R.A.S., F.C.S., Trinity College, Dublin, gave the following results :—

Density of fresh juice	1·040
Ditto after 24 hours' exposure to air	1·043
100 parts of juice by weight, yielded of	
Sugar	10·700
Tannin, Mucilage, Salts, &c.	2·380
Water	86·920

The tree grows to a large size, and bears an abundance of highly coloured fruit.

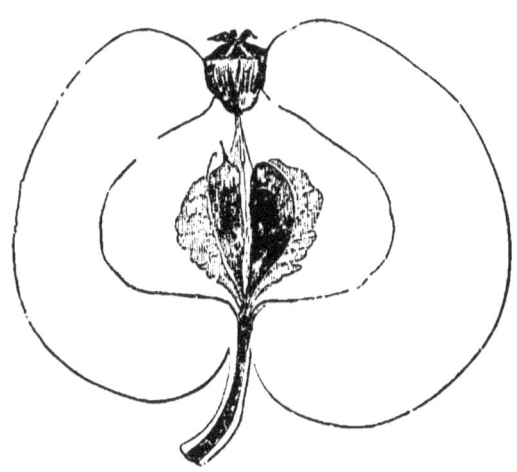

BROMLEY.

A very old variety, spread throughout the orchards of Gloucestershire and Herefordshire, but not abundant in the latter county. Its history is lost.

Description.—Fruit: middle sized, roundish, and flattened, very uneven, and angular on the sides; and knobbed both at the crown and base. Skin: bright yellow, much covered with firm broken streaks of crimson, nearly over the whole surface, but especially where exposed to the sun; russety all over the base, whence it extends in lines up the sides. Eye: closed, with broad, flat, convergent segments, set in a deep angular basin; tube, funnel shaped; stamens, basal. Stalk: straight, and stout, from half to three quarters of an inch long, set in a deep cavity. Flesh: yellowish, firm, and somewhat woolley in texture. Juice: pale, plentiful, fairly sweet, with a brisk acidity. Cells of the core, open.

The chemical analysis of the juice of the *Bromley* apple (season 1880), by Mr. G. H. With, F.R.A.S., F.C.S., Trinity College, Dublin, gave the following results:—

Density of fresh juice	1·033
Ditto after 24 hours' exposure to air ...	1·035
100 parts of juice by weight, yielded of	
Sugar	12·100
Tannin, Mucilage, Salts, &c.	1·300
Water	86·600

This analysis does not indicate any high merit, though the the apple is still held in great esteem in Gloucestershire, where it is thought nearly equal to *Skyrme's Kernel*. The cider is said to be strong, but not sweet. It is good for cooking, and as an apple for sauce is unsurpassed. It is a late apple, keeps well, and sells well in the market; all merits, that help no doubt to keep it in favour.

The tree grows to a large size, spreading broadly. It is shy in bearing, and has not been much propagated of late years.

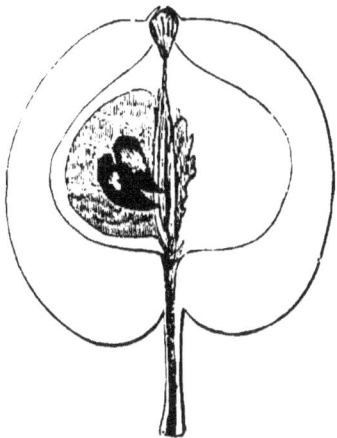

CARRION APPLE.

[SYN: *Kempley Red.*]

This variety takes its name, in the Pyon district of Herefordshire, from the fact of one of the oldest trees being used to hang the dog's meat on. It has been exhibited in the Hereford Apple Shows, under the name of *Kempley Red*.

Description.—Fruit : small, oblong, and regular in shape. Skin : with a yellowish green coloured ground in the shade, but the whole surface is nearly covered with crimson, which becomes very dark, on the side next the sun, with splashes of a deeper shade all over the fruit. Eye : small and closed, level with the surface. Stalk : long and slender, inserted in a very small and narrow cavity. Flesh : yellow, pink tinted near the skin. Juice : small in quantity, of full amber colour, viscid, sweet, with some astringency.

The chemical analysis of the juice of the *Carrion Apple* (season 1883), by Mr. G. H. With, F.R.A.S., F.C.S., Trinity College, Dublin, gave the following results :—

Density of fresh juice	1·050
Ditto after 24 hours' exposure to air	1·050
100 parts of juice by weight, yielded of	
Sugar	12·800
Tannin, Mucilage, Salts, &c.	1·500
Water	85·700

The tree grows to a medium size, and is very prolific.

CHERRY HEREFORD.

[Syn : *Cherry Norman; Hitterly.*]

The history of this apple is not known. It is much grown about Marden, and other places in the valley of the Lugg.

Description.—Fruit : round, pretty regular in outline, sometimes a little ribbed at the sides, but very round at the base, with a small and very narrow stalk cavity. Skin : clear straw yellow, with a russet cheek on the sunny side, and a dash of crimson, or orange red; the russet extends in tracings to the shady side. Eye : very small, and placed in a shallow depression, set round with prominent plaits; segments, convergent; tube, conical; stamens, marginal. Flesh : soft, spongy, slightly bitter and sweet. Cells of the core, slightly open; cell-walls, roundish obovate.

The chemical analysis of the juice of the *Cherry Hereford*, by Mr. G. H. With, F.R.A.S., F C.S , Trinity College, Dublin, gave the following results:—

Density of fresh juice ...	1·043
Ditto after 24 hours' exposure to air	1·046
100 parts of juice by weight, yielded of	
Sugar	12·830
Tannin, Mucilage, Salts, &c.	2·073
Water	85·097

The *Cherry Hereford* is one of the best early fruits. It makes cider of a deep colour, with a sweet, rich, and pleasant flavour.

The tree grows well and freely; it blossoms in the middle of May, and ripens its fruit in the middle of October. It is apt to bear in abundance, only every second year.

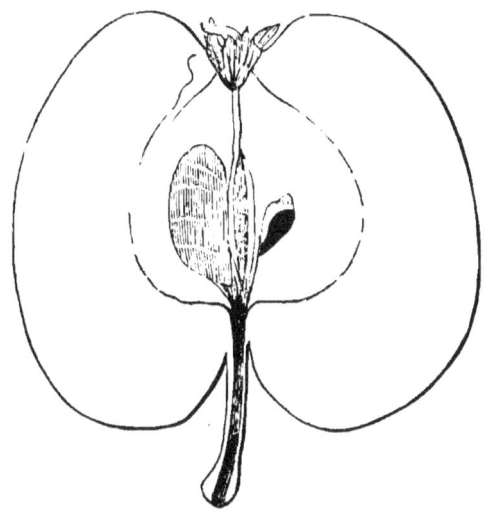

CHERRY PEARMAIN.

This variety is an old one, and widely spread throughout the orchards of Herefordshire. It is without any known history.

Description.—Fruit: very handsome in colour, and regular in shape, round oblong, above medium size. Skin: yellow, but very much covered, as it were, with the small particles of broken-up streaks of crimson, which run together on the side exposed to the sun, where they are traversed by streaks of deeper crimson. Eye: small and closed, set in a narrow cavity. Stalk: half an inch long, almost hidden in a deep and narrow cavity, which is lined with russet. Flesh: soft, and reddish pink in patches beneath the skin, and outside the core fibres; sweet and pleasant to taste, with a slight after roughness. Juice: plentiful, of a rich pink colour, changing to a deep rosy red.

The chemical analysis of the juice of the *Cherry Pearmain*

(season 1882), by Mr. G. H. With, F.R.A.S., F.C.S., Trinity College, Dublin, gave the following results :—

Density of fresh juice	1·047
Ditto after 24 hours' exposure to air	1·050
100 parts of juice by weight, yielded of	
Sugar	12·700
Tannin, Mucilage, Salts, &c.	2·000
Water	85·300

A very favourite apple in the orchard. It is handsome in shape, and colour. It is very good, eaten fresh from the tree; will make a pudding; or mix well with other varieties in the cider vat.

The tree is of good size, and generally hardy, but in some localities it is apt to canker. It bears freely.

There is a red variety of *Cherry Pearmain*, which differs but little from this one, except in its deeper colour.

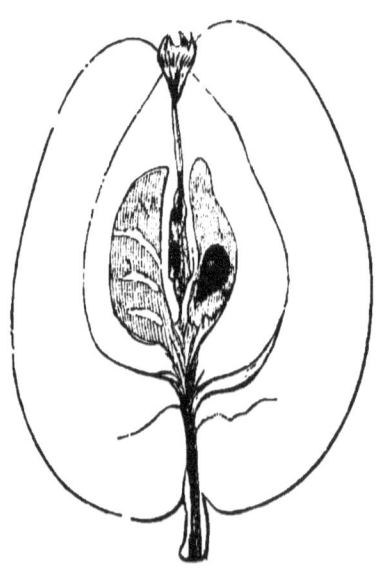

CIDER LADY'S FINGER.

The origin of this variety, does not seem to be known, but

from the age of the trees, it was probably produced at the end of the last, or the beginning of the present century.

Description.—Fruit : of middle size, two and a half inches long, by one inch and a half wide, oblong, even, but not always regular in its outline, with a waist near the top. Skin : quite smooth, dull orange, or yellow on the shaded side, with a few broken stripes of red ; washed with thin red, which is streaked with darker and brighter red, on the side next the sun ; the whole surface being strewed with russet specks. Eye : small, and prominently set ; open, with very short divergent segments, and surrounded with a few prominent plaits, or little knobs ; tube, funnel shaped ; stamens, marginal. Stalk : very slender, short, inserted in a shallow cavity, or merely in a slight depression, surrounded with russet. Flesh : yellowish, rather dry ; juice of a fine rich colour, with a sweet, sub-acid and astringent flavour. Cells of the core, open.

The chemical analysis of the juice of the *Cider Lady's Finger* (season 1878), by Mr. G. H. With, F.R.A.S., F.C.S., Trinity College, Dublin, gave the following results :—

Density of fresh juice ...	1·041
Ditto after 24 hours' exposure to air	1·045
100 parts of juice by weight, yielded of	
Sugar	13·242
Tannin, Mucilage, Salts, &c. ...	1·412
Water ...	85·346

This apple is a valuable addition to the orchard. It ripens early, but it is easy to manage, and makes very good cider. It is rich, strong, and brisk, often good enough to bottle ; but is apt in the hot weather of early autumn, to lose much of its richness from over fermentation.

The tree is hardy ; it blossoms in the beginning or middle of May, and bears a profusion of fruit, which ripens in September. It is a variety growing into favour, and deservedly so. It is much grown in the orchards of the River Froome Valley, and is becoming widely distributed throughout Herefordshire.

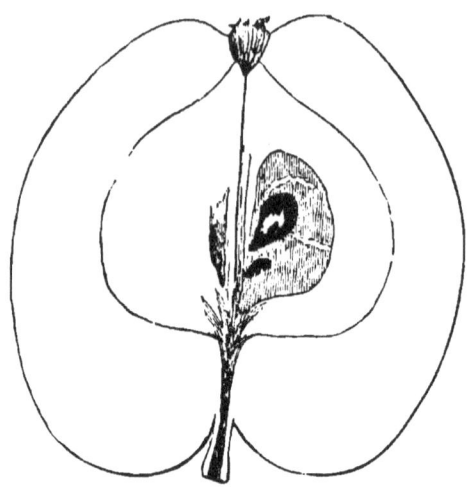

COCCAGEE.

[SYN : *Cocko Gee ; Cockagee.*]

A very old variety, believed to be of Irish origin, but its real history seems lost. It is said to have been "brought into Somersetshire by Counsellor Pyne, a gentleman, who resided near Exeter, and who had the care of Mr. William Courtenay's Estates in Ireland." *Treatise on Cyder Making*, by Hugh Stafford, of Pynes (1753).

Description.—Fruit : medium size, very variable in shape, but usually ovate. Skin : yellow in the shade, marked with green specks, with a deep blush of red next the sun. A reddish tint is often spread over the whole fruit, and not unfrequently, patches of thin russet. Eye : small and closed, set in a narrow plaited basin. Stalk : short, inserted in a narrow but rather deep cavity, frequently lined with russet. Flesh : yellowish white, firm and crisp. Juice : moderately plentiful, of an amber colour, and a harsh austere taste.

A very old, and highly esteemed variety for culinary purposes, and especially for baking, when it possesses a peculiarly rich flavour. "This apple" says Brookshaw, "triumphs over all others in sauce, tarts and pies, as much as its juice does in cider. No cook would ever make use of any other apple if he could get this. It is so extremely rough and tart, that it would be almost impossible to eat one raw." It is in season from October, to February and March.

The chemical analysis of the juice of the *Coccagee* (season 1880), by Mr. G. H. With, F.R.A.S., F.C.S., Trinity College, Dublin, gave the following results :—

Density of fresh juice	1·052
Ditto after 24 hours' exposure to air	1·058
100 parts of juice by weight, yielded of	
Sugar	9·080
Tannin, Mucilage, Salts, &c.	7·820
Water	83·100

As a cider fruit, it has long possessed the highest repute. The *Coccagee* apple was the favourite cider apple of Devonshire, at the beginning of the present century. There was a celebrated orchard at Heathfield, near Milverton, from which it is said that the cider was supplied by the Rev. Mr. Cornish, for the Queen's household, at £10 10s. the hogshead. "I find nothing extraordinary in it;" says Mr. Stafford in his book on Cyder Making (1753), "'tis true it has a vinous pipinary golden flavour;" and so authorities differ. The analysis, however, proves its merit.

The tree is very hardy, and bears well, but it has not been much cultivated of late years.

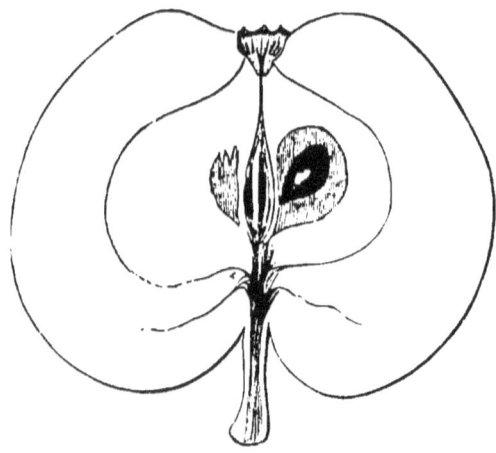

COWARNE RED.

This fruit takes its name from the parish of Much Cowarne, Herefordshire, where it was raised about the beginning of the last century, (c. 1720). This apple is well represented in the "Pomona Herefordiensis," Plate xxviii.

Description.—Fruit: above medium size, roundish oblate, narrowing towards the crown, where it has a few obtuse ribs more or less defined. Skin: golden yellow on the shaded side, with numerous streaks of red, a bright red over almost all the surface, and where fully exposed to the sun, becoming of a deep purplish crimson. Eye: small and closed, and set in a narrow cavity. Stalk: half an inch long, stiff and straight, deeply inserted in a narrow cavity which is lined with very thin russet. Flesh: crisp and pleasant to taste, tinted with crimson beneath the skin and slightly marking the fibre of the core. Juice: very thin and plentiful, of a ruddy amber colour, and very slightly astringent.

The chemical analysis of the juice of the *Cowarne Red* (season 1882), by Mr. G. H. With, F.R.A.S., F.C.S., Trinity College, Dublin, gave the following results:—

Density of fresh juice	1·047
Ditto after 24 hours' exposure to air	1·047
100 parts of juice by weight, yielded of	
Sugar	11·900
Tannin, Mucilage, Salts, &c.	1·400
Water	86·700

Mr. Knight makes the specific gravity of this apple as high as 1·069.

The *Cowarne Red* is a favorite variety in the orchard. It is a good apple, but its bright colour, and its free bearing habit has certainly helped its popularity.

The tree grows to a large size, is very hardy, and is often to be seen in extreme old age.

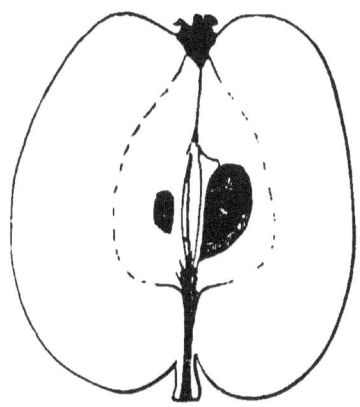

CUMMY.

[SYN: *Cummy Norman.*]

This variety has no published history, but it is believed to have been raised at Cummy, in Radnorshire.

Description.—Fruit : conical, even and regular, except when it has occasionally one or two rather prominent angles on the sides; wide at the base, and very narrow at the apex. Skin: greenish yellow on the shaded side, and with a thin, red cheek, speckled with deep crimson on the side next the sun, and sprinkled over the surface with minute russet dots. Eye : prominent, closed, set in a narrow, plaited basin, segments, broad and leaf-like. Stalk : a quarter of an inch long, slender, inserted in a close, deep, irregular cavity, which is lined with russet. Flesh : very tender, juicy, and with a slight aromatic, bittersweet flavour, without astringency. The juice is of a deep amber colour.

The chemical analysis of the juice of the *Cummy* (season

1881), by Mr. G. H. With, F.R.A.S., F.C.S., Trinity College, Dublin, gave the following results:—

Density of fresh juice	1·033
Ditto after 24 hours' exposure to air	1·040
100 parts of juice by weight, yielded of	
Sugar	14·000
Tannin, Mucilage, Salts, &c.	·060
Water	85·940

The abundance of saccharine matter contained in this fruit justifies the general esteem in which it is held. It does not make cider of the first quality when used alone, but it gives body and strength to other varieties, and they must supply flavour and good keeping qualities.

The tree grows freely and is very hardy. It blossoms and bears profusely year after year in almost any situation.

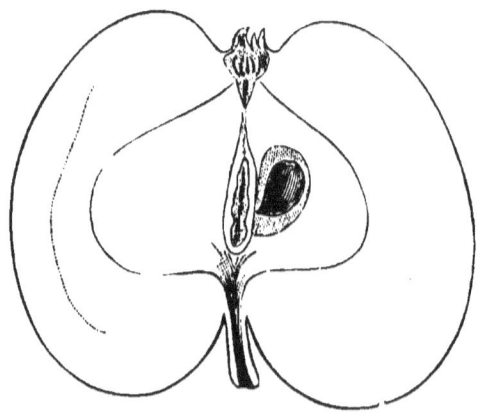

DE BOUTTEVILLE.

A seedling raised at Yvetot, by Monsieur Legrand. It first fruited in 1873, and was dedicated to Monsieur L. De Boutteville, Honorary President of the SOCIÉTÉ CENTRALE D'HORTICULTURE

DE LA SEINE INFÉRIEURE, and Author, with Monsieur A. Hauchecorne, of the celebrated work *Le Cidre*, published at Rouen, in 1875. This variety was introduced into Herefordshire by the Woolhope Club, in 1884.

Description.—Fruit: of middle size, oblate, smooth and round, without angles. Skin: pale yellow, with an orange blush on the sunny side, more or less spotted over the surface, and the spots often become dark and tinged with red under the sun's influence. Eye: closed, seated in a narrow, deep cavity, with folded margins. Stalk: short, placed in a broad and deep cavity, lined with a thin russet that radiates over the base of the apple. Flesh: yellowish, with a sweet and pleasant flavour, free from bittterness. Juice: of a high colour, sweet, and pleasant.

"This apple," says Monsieur Hauchecorne, "is one of the best varieties for making a good cider that will keep well. The apple is firm in flesh, and travels well. Its juice is well coloured with excellent perfume and taste." It has a density of 1,083. One thousand parts contain of alcoholic sugar 193; tannin 6; mucilage 11; acidity 2.14; salts, &c., 7.86; and water 780.

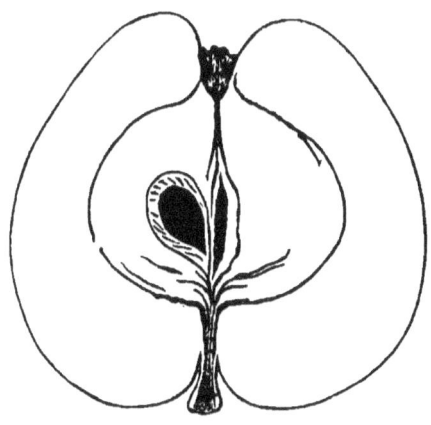

DYMOCK RED.

This apple takes its name from the village of Dymock, in

Gloucestershire, on the borders of Herefordshire. It is an apple of considerable antiquity, and was probably produced towards the end of the seventeenth century. In Evelyn's time it bore a high reputation, and it well sustains its character in these days.

Description.—Fruit: roundish or oblate, even and regular in its outline; handsome. Skin: entirely covered with dark mahogany red, with streaks of bright pale crimson on the side next the sun, and somewhat paler, though of the same colour, on the shaded side; the whole surface is strewed with distinct russet dots, and mottled with patches, and ramifications of cinnamon coloured russet. Eye: medium sized, with segments that are sometimes divergent and sometimes connivent; when the former, they are quite reflexed, and when the latter, they touch each other by their margins and close the eye, which is placed in a narrow, slightly plaited basin; tube, funnel shaped; stamens, basal; stalk, very short, and often a mere knob, in a very narrow and shallow cavity. Flesh: yellowish, tender and soft, occasionally tinged with red, slightly sweet, with a pleasant acidity. Cells of the core, closed; cell-walls, ovate.

The chemical analysis of the juice of the *Dymock Red* (season 1878), by Mr. G. H. With, F.R.A.S., F.C.S., Trinity College, Dublin, gave the following results:—

Density of fresh juice ...	1·033
Ditto after 24 hours' exposure to air	1·037
100 parts of juice by weight, yielded of	
Sugar	12·100
Tannin, Mucilage, Salts, &c.	3·280
Water ...	84·620

The cider made from this apple, whether pure, or mixed with other fruit, is rich and excellent.

The *Dymock Red* apple is chiefly grown in the neighbourhood of Ledbury, but from its high merits it deserves a far wider cultivation. The colour of the apple is a deep dull red, in sunny seasons it takes quite a mahogany tint.

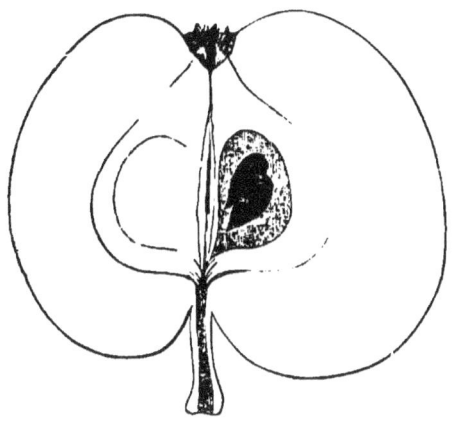

EGGLETON STYRE.

This apple was raised from the kernel by the late Mr. William Hill, at Lower Eggleton, Ledbury, Herefordshire, in the nursery attached to the farm. The seedling first bore fruit about the year 1847, and it was from the birds specially attacking the apple, that Mr. Hill's attention was directed to their sweet and rich flavour.

Description.—Fruit: middle sized, roundish, with obscure ribs on the sides. Skin: rich yellow, orange next the sun, and covered with thin tracings and patches of russet. Eye: open, with reflex segments like *Court of Wick*, set in an even basin. Tube: short and funnel shaped; stamens, medium. Stalk: slender, half an inch long, deeply inserted in a round cavity, which is lined with russet extending in branches over the base. Flesh: yellowish, tender, juicy, sweet and slightly acid. Cells of the core, open.

The chemical analysis of the juice of the *Eggleton Styre* (season 1880), by Mr. G. H. With, F.R.A.S., F.C.S., Trinity College, Dublin, gave the following results :—

Density of fresh juice ...	1·049
Ditto, after 24 hours' exposure to air	1·050
100 parts of juice by weight, yielded of	
Sugar	10·591
Tannin, Mucilage, Salts, &c. ...	6·569
Water ...	82·840

The *Eggleton Styre* makes excellent cider alone, very sweet and rich, with a high colour. It has been sold, fresh bottled, at 16/- the dozen. It fines better if mixed with *Redstreak*, *Cowarne Red*, *Pym Square*, *Cook's Kernel*, or *Strawberry Hereford*.

The tree is hardy. It blossoms the middle of May, and bears freely and ripens its fruit in October. The fruit is so sweet and aromatic as to be very attractive to hares, rabbits, fowls, blackbirds, and fieldfares, not to mention smaller birds. They will select this variety in preference to all others.

The *Eggleton Styre* is chiefly grown in the parish of Eggleton and the surrounding orchards, but it is gradually spreading throughout the county.

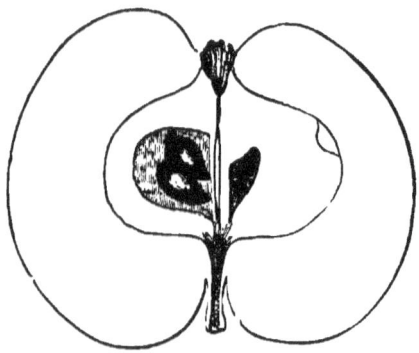

FOREST STYRE.

[SYN: *Stire; Stirom.*]

A fine old Gloucestershire cider apple, extensively cultivated on the thin light limestone soil of the Forest of Dean. Its origin is lost. It is mentioned by Philips the cider poet.

> "*Stirom* firmest Fruit,
> Embottled (long as *Priameian Troy*
> Withstood the *Greeks*) endures, e'er justly mild.
> Softened by age, it youthful Vigour gains,
> Fallacious drink! Ye honest men beware!
> Philips "*Cyder*."

Description.—Fruit: below medium size, roundish, inclining to oblate, regularly and handsomely shaped. Skin: pale yellow, with a blush of orange on the side exposed to the sun, and numerous small russet spots, scattered over the surface. Eye: small and closed, with short obtuse segments, and set in a narrow basin, more or less plaited. Stalk: short, in a narrow deep cavity, lined with russet throughout, and which spreads from it over the base of the apple. Flesh: yellow, dry and harsh. Juice: small in quantity, pale straw colour, changing to deep amber, with a remarkably sweet, luscious flavour and some astringency.

The chemical analysis of the juice of the *Forest Styre*, by Mr. G. H. With, F.R.A.S., F.C.S., Trinity College, Dublin, gave the following results:—

Density of fresh juice ...	1·073
Ditto after 24 hours' exposure to air	1·074
100 parts of juice by weight, yielded of	
Sugar	14·000
Tannin, Mucilage, Salts, &c.	3·300
Water ...	82·700

"The Forest Stire," says Mr. Thomas Andrew Knight, "is almost universally supposed to afford a stronger cider than any other kind of apple." He found its specific gravity to be as high "as 1·076 to 1·081 according to the soil it grows in."

The trees grow with numerous upright shoots, like a pollard willow, and are not renowned for bearing well. Marshall, in his "Rural Economy" (1796), speaks of this variety as decaying rapidly.

There are very few trees left in Gloucestershire at this time, but it will be seen on reference to page 26 of this work, that Mr. William Viner Ellis, of Minsterworth, has sent apples and grafts of this valuable variety to the Woolhope Club, and that Messrs. Cranston and Co., of King's Acre, near Hereford, have succeeded in propagating it.

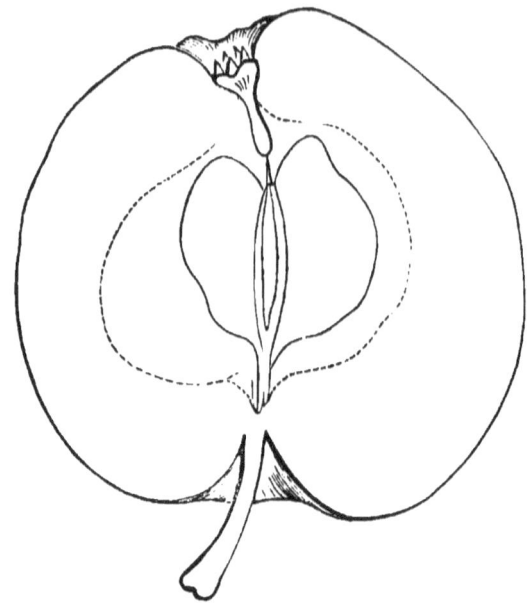

THE FOXWHELP APPLE.

> "Cider for strength and a
> long-lasting drink is best
> made of the Foxwhelp of the
> Forest of Deane, but which
> comes not to be drunk till
> two or three years old."
> *(Appendix to Evelyn's "Pomona." Edit.* 1706*).*

The Foxwhelp apple is the favourite cider apple of Herefordshire. Its origin and its singular name are alike obscure.

The earliest record we have of the *Foxwhelp* is by Evelyn in his "*Pomona*," which is an Appendix to the *Sylva* "concerning fruit trees in relation to cider." This was first published in 1664, and at that time and long after, the great Apple of Herefordshire was the *Redstreak*. The *Foxwhelp* is disposed of in a few words— "Some commend the *Foxwhelp*." Ralph Austen, who wrote in 1653, makes no mention of it when he says, "Let the greatest number of fruit trees, not onely in the orchards but also in the fields be *Pearmaines, Pippins, Gennet-Moyles, Redstreaks,* and such kinds as are knowne by much experience to be especiall good for cider." Neither is any notice taken of it by Dr. Beale in his "*Herefordshire Orchards*, written in an epistolary address to Samuel Hartlib, Esq.," in 1656.

The first notice of it, after Evelyn, is by Worledge in 1676, who merely says, "The *Foxwhelp* is esteemed among the choice cider fruits." In Evelyn's time it seems to have been regarded as a native of Gloucestershire, for Dr. Smith in the "*Pomona*" when writing of "the best fruit" (with us in Gloucestershire) says, "the cider of the *Bromsbury Crab* and *Foxwhelp* is not fit for drinking till the second year, but then very good"; and in the quotation at the head of this paper "A person of great experience" calls it "the *Foxwhelp* of the Forest of Deane."

Its great merit as a cider apple seems to have been quickly recognised, but its cultivation up to this period could not have been on an extensive scale, or it would have been more generally known. Even Philips in his celebrated poem, entitled "*Cyder*" seems as ignorant of its existence, as most of the writers on orchards were at that period. A highly appreciative notice of it is found in a letter to a friend, written by Hugh Stafford, of Pynes in Devonshire, Esq., bearing date 1727. He says,. "This is an apple long known, and of late years has acquired a much greater reputation than it had formerly. The fruit is rather small than middle-sized; in shape long, and all over of a dark red colour. I have been told by a person of credit, that a hogshead of cider from this fruit has been sold in London for £8 or eight guineas, and that often a hogshead of French wine, has been given in exchange for the same quantity of *Foxwhelp*. It is said to contain a richer and more cordial juice than even the *Redstreak* itself, though something rougher if not softened by racking. The tree seems to want the same helps as the *Redstreak* to make it grow large. It is of Herefordshire extraction." Mr. Knight in the "Pomona Herefordiensis," published in 1811, also thought it "certainly a true Herefordshire apple," and this of late, has been the prevalent belief, derived probably from the opinion of the two last named writers.

The merit of its production thus rests with the *Forest of Dean*, on the authorities we have given, but there is no record of the origin of its singular name. It may readily be supposed, however, that the stray seedling sprang up near a fox's earth, and thus, when it had shown its character, it obtained its name. Some devoted admirers think they see in the eye of this apple, a distinctive resemblance to the physiogonomy of a young fox, but here, surely the name has guided the imagination. Wherever the young seedling may have grown, the brilliant colour of its fruit would render it conspicuous, and its rough peculiar flavour, with a judge of apples, would proclaim its merit. It is probable, that a fox-hunter found and named it, and certainly none appreciate more highly than fox-hunters, the merits of its cider.

Description.—The fruit is roundish, inclining to conical or ovate, with an uneven outline, caused by several obtuse ribs on the sides, and which terminate in ridges round the eye; in good specimens one side is generally convex, while the other is flattened. Skin: beautifully striped with deep bright crimson and yellow; on the side next the sun it is more crimson than it is on the shaded side, where the yellow stripes are more apparent. Eye very small, set in a narrow, shallow, and plaited basin; segments short, somewhat erect, and slightly divergent. Calyx-tube, funnel shaped. Stamens, marginal. Stalk, three-quarters of an inch long, obliquely inserted by the side of a fleshy swelling, which pushes it on one side and gives it a curving direction. Flesh: yellow, tinged with red, tender, and with a rough and acid flavour. Cells of the core, wide open. It belongs to group 10 of Dr. Hogg's New Classification of Apples.

The surface of the *Foxwhelp* apple is usually marked by small dark coloured circular scabs or patches, which are thought by some growers to be characteristic of the *Foxwhelp*, but this is not so. The round patches are formed by the miscroscopic fungus, *Spilocæa pomi*, and are commonly to be found on the apples of very aged trees, of all kinds of fruit. Like all fungus growths, this is much more abundant in some seasons than in others.

The coloured plate of the *Foxwhelp* apple in the Herefordshire Pomona, was drawn from fruit grown on the estate of W. H. Apperley, Esq., of Withington. The trees are believed to have been planted by one of his ancestors, about the year 1609, and are still in fruitful vigour.

The form of the fruit varies according to the age of the tree, and this is the case with most varieties. The section is taken from fruit grown by John Bosley, Esq., of Lyde, and represents a fruit from a tree which is the result of four successive graftings, from one of the old trees of the *Foxwhelp*, the scions being taken in each instance from the tree grafted the previous year.

A *Foxwhelp* apple of good size and colour, grown in the year 1876, yielded 7½ drachms of a strongly acidulated juice with its own flavour, and of the specific gravity of 1·068; and others of a smaller size gave 5½ drachms of juice with a specific gravity of 1·074. Mr. Knight gives the higher specific gravity of 1·076 to 1·080, which perhaps might be due to a more favourable year.

The chemical analysis of the juice of the *Foxwhelp* (season 1877), by Mr. G. H. With, F.R.A.S., F.C.S., Trinity College, Dublin, gave the following results:—

Density of fresh juice ...	1·068
Ditto after 24 hours' exposure to air	1·070
100 parts of juice by weight, yielded of	
Sugar	14·400
Tannin, Mucilage, Salts, &c. ...	8·500
Water	77·100

It must be stated, however, that the absence of sun, and the great rainfall of the summer of 1877, made it a most unfavourable season for the growth of any fruit in perfection.

The home of the *Foxwhelp Apple*, be its origin what it may, is in the deep clay loam of the Old Red Sandstone, in the central districts of Herefordshire, and especially in the valleys of the rivers Lugg and Froome. The chief orchards are to be found in the villages of Lugwardine, Westhide, Withington, Lyde, Moreton, Sutton, Wistaston, Marden, Bodenham, Burrop, Wellington-on-the-Lugg; and those of Weston Beggard, Yarkhill, Tarrington, Stoke Edith, Stretton Grandison, Eggleton, the Froomes, the Cowarnes, and the other villages on the Froome, are seldom without a few old trees of the *Foxwhelp Apple*.

The broad valley of the Wye does not generally present so good and rich a soil. The river has been so erratic in days gone by, that large beds of gravel and marl are to be met with in all directions, and the orchards of repute therefore are only to be found on the rising slopes of the valley, out of the river's reach. There are many excellent orchards from King's Caple and Holme Lacy by Credenhill to Kinnersley, Sarnesfield, Dilwyn, and the Weobley district; the *Foxwhelp* may be found in any of them, and wherever it is found, it is treasured greatly for its valuable fruit.

The *Foxwhelp Apple* tree is upright and handsome in growth, where age has not rendered it rugged and gnarled. It is a slow growing tree, and a shy, capricious bearer, and this may perhaps partly explain, why fruit growers should prefer to propagate those sorts which grow more freely, and are more certain croppers. The tree is hardy, and its fruit is in great demand. There is yet a want of young trees generally, for, be the reason what it may, grafts of late years have not succeeded well. The orchardists however have only to apply themselves to the cultivation of the *Foxwhelp*, and resolutely determine to perpetuate this precious variety, and the same success will crown their efforts in the future, which followed those of their predecessors in the past.

The *Foxwhelp* cider, when pure, is of great strength, and always has a peculiar aroma, so marked that it can be detected directly the cork is drawn from the bottle. In taste, it is generally rough and strong, with a peculiar vinous, musky flavour, which gives its aroma. In ordinary seasons, unless made with great care, it is not sweet enough to be acceptable to strangers, and the taste which enjoys its peculiar flavour fully, must in such circumstances, perhaps, be acquired; but in a favourable year—a year of sunshine and genial showers, when the fruit has been ripened to perfection—happy is he who has a good hit of it. If he carries it well through the process of fermentation, and keeps the flavour of the fruit, and its sweetness too, he has cider in perfection—a cider that will sell readily in its own district, at a guinea a dozen; and a cider moreover, that will unquestionably improve in quality, for some three or four decades of years. It will not all be sold, however, for it is the pleasure and pride of the cider-growers of Herefordshire to have always ready for a friend, a bottle of good *Foxwhelp* cider of a good year.

The juice of the *Foxwhelp Apple* is, however, most used to give strength and flavour to the cider of mixed fruit, and when this is well made, it is perhaps more generally popular than the very strong and pure *Foxwhelp*. A cider of this kind, excellent in quality, can be got at one shilling a bottle from the growers. The *Foxwhelp* cider has the character of changing colour very quickly, on exposure to the air, and even at the table, if not drunk quickly, the dusky greenish tint will show itself. Some other strong ciders have also this peculiarity, which is certainly not a virtue.

The *Foxwhelp*, beyond all question, in general estimation is the most valuable cider apple, and by intelligent perseverance in propagating it, it will long continue to be so.

The Woolhope Club, it will be seen on reference to page 26 of this work, has succeeded in propagating it extensively, and will thus have rescued this valuable variety from loss.

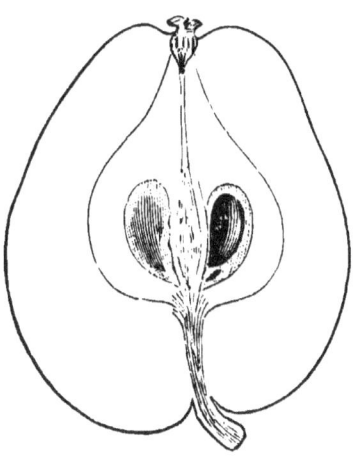

FRÉQUIN AUDIÈVRE.

A seedling raised by Monsieur Audièvre, treasurer of the SOCIÉTÉ D'HORTICULTURE D'YVETOT, in 1868. It is thought to have been a seedling from *Petit-Fréquin* or *Fréquin Rouge*, with greatly improved qualities to either of these varieties. It was introduced into Herefordshire, in 1884, by the Woolhope Naturalists' Field Club.

Description.—Fruit: very small, flattened at the base, but contracting rapidly towards the eye. Skin: with a pale yellow ground, almost entirely covered with red carmine, and frequently with many fine white spots on the surface. Eye: small and closed, set in a narrow cavity with sulcated borders. Stalk: variable, generally very short, and set obliquely in a small and shallow cavity. Flesh: yellowish white, and firm. Juice: sweet, slightly bitter, but with good perfume and flavour.

"This valuable variety," says Monsieur Hauchecorne, "possesses the highest merit of the *Fréquin* tribe. It contains all the elements for making a strong, pleasant, and healthy cider." The juice has a very high colour, and a density of 1·079. One thousand parts contain of alcoholic sugar 180; tannin 5·509; mucilage 12; acidity 1·320; salts, &c., 11·171; and water 790.

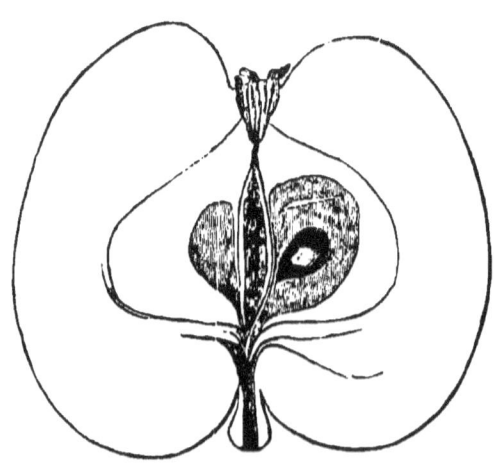

GARTER APPLE.

This variety is spoken of by Mr. Thomas Andrew Knight, at the beginning of the century, as a comparatively new apple "much cultivated during the decay of the older, and more valuable varieties." It is figured on Plate xxvi., in the "Pomona Herefordiensis."

Description.—Fruit: full medium size, two and a half inches high, and the same measurement in breadth, smooth and evenly shaped, broad at the base and tapering towards the crown. Skin: smooth, yellow in the shade, with a warm crimson blush on the side next the sun, and broken streaks of a much deeper colour, spotted over with numerous very small dark, or red specks. Eye: small and open, deeply sunk in a narrow cavity, which is lined with very pale russet. Flesh: soft and white. Juice: moderate in quantity, of a rich amber colour, sweet, subacid, astringent, and rich in flavour.

The chemical analysis of the juice of the *Garter Apple* (season 1883), by Mr. G. H. With, F.R.A.S., F.C.S., Trinity College, Dublin, gave the following results:—

Density of fresh juice	1·063
Ditto after 24 hours' exposure to air	1·064
100 parts of juice by weight, yielded of	
Sugar	12·540
Tannin, Mucilage, Salts, &c.	2·260
Water	85·200

A favourite fruit in the Herefordshire orchards. Very attractive in colour. It is used as a dessert as well as a cider fruit. Mr. Knight made the specific gravity of the juice to be 1·066 in a favourable season.

The tree grows freely, is very hardy, and bears abundantly.

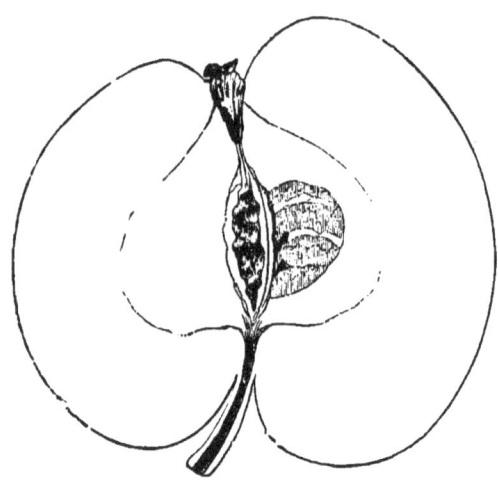

GENNET MOYLE.

> The Moile
> Of sweetest honey'd taste.
> *Philips Cyder.*

The *Gennet Moyle* was the favourite apple in the Cider Orchards of the 15th century, and continued to be so until Lord Scudamore's *Redstreak* supplanted it in popular esteem. Its history is lost; but its name signifies "a hybrid scion," from "gennet" a hybrid, or mule; and "moyle" a scion, or graft. It is still to be found in the old orchards of Herefordshire, though it has now become scarce.

Description.—Fruit: round, somewhat prominently and obtusely ribbed on the sides, and with ridges round the crown. Skin: of a clear lemon colour, with more or less russety cheek, and with russet lines all over the side exposed to the sun. Eye: closed, with convergent, leafy segments, and set in a puckered basin; tube, long and funnel shaped; stamens, marginal. Stalk: about half an inch long, inserted all its length in the cavity, which is lined with russet. Flesh: with a yellowish tinge, tender, not very juicy, but rather dry, and with a very sweet, slightly acid flavour. Cells of the core, open.

The chemical analysis of the juice of the *Gennet Moyle* (season 1880), by Mr. G. H. With, F.R.A.S., F.C.S., Trinity College, Dublin, gave the following results :—

Density of fresh juice ...	1·046
Ditto after 24 hours' exposure to air ...	1·053
100 parts of juice by weight, yielded of	
Sugar	9·570
Tannin, Mucilage, Salts, &c.	5·430
Water	85·000

This sweet and fragrant apple is now very scarce. Reference must be made to the old writers for its character. Dr. Beale says of it "our *Gennet Moyles* are commonly found in hedges, or in our worst soil, most commonly in *Irchenfield*, or towards Wales, where the land is somewhat dry and shallow. This fruit is nice and apt to be discouraged by blasts, and we do ordinarily expect a failing of them every other year. But this fruit makes the best *Cyder* in my Judgment, and such as I do prefer before the much commended *Redstreak'd*. For this *Gennet Moyle* if it be suffered to ripen on the Tree, and not to be mellow, but to be yellowish and fragrant, and then to be hoarded in Heaps under Trees, a fortnight or three Weeks before you grind them; it is (at a distance) the most fragrant of all Cyder Fruit, and gives the Liquor a most delicate perfume. So for Tarts and Pyes it is much commended." *Herefordshire Orchards* (1730).

In Evelyn's *Pomona*, the *Gennet Moyle* of one year is named first as a Summer Cyder, and of the fruit it is added "The best Baking apple that grows; and it keeps long, baked; but not so, unbaked, without growing mealy. It dries well in the oven, and with little trouble."

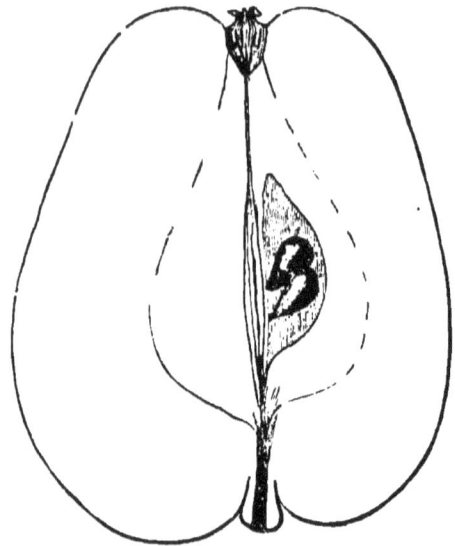

GREEN WILDING.

This variety is without history, and is probably a seedling from some small orchard nursery. From the age of the original tree it must be as old as this century.

Description.—Fruit: middle size, conical, obscurely ribbed, narrowing towards the eye, where it is somewhat puckered. Skin: yellowish green, strewed with numerous large, russety dots, and a few lines of russet. Eye: small, and closed, set in a narrow puckered basin. Stalk: very short, completely embedded in a deep cavity. Flesh: white, tender, sweet, and with a mawkish flavour, but without either bitterness, marked astringency, or much acidity.

The chemical analysis of the juice of the *Green Wilding* (season 1881), by Mr. G. H. With, F.R.A.S., F.C.S., Trinity

College, Dublin, gave the following results:—

Density of fresh juice	1·044
Ditto after 24 hours' exposure to air	1·046
100 parts of juice by weight, yielded of	
Sugar	10·530
Tannin, Mucilage, Salts, &c.	3·170
Water	86·300

The *Green Wilding* grows in the valley of the river Froome, where it is highly esteemed. It makes a good, sound, deep coloured cider, with a sweet and pleasant flavour, but it is generally mixed with other varieties. The analysis proves it an excellent fruit.

The tree is upright in growth, very hardy, and a good bearer. It is grown chiefly about Eggleton, Homend, and in the adjoining orchards, where many large trees are to be found. It is still being propagated on that side of the county.

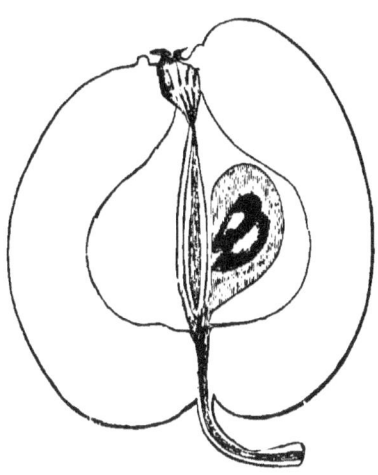

HAGLOE CRAB.

This fruit originated at Hagloe, in the parish of Awre, Gloucestershire, and was first brought into notice by Mr. Bellamy who lived there. Marshall in his "Rural Economy of Gloucestershire," states that it was raised from seed about the year 1720, but Mr. Thomas Andrew Knight in the "Pomona Herefordiensis," 1811, thinks that the excellence of the apple was only then first discovered, for his friends had sought in vain many years before, for the original tree at Awre. A coloured illustration is given by Mr. Knight, Plate V., of this apple.

Description.—Fruit: small, ovate, narrowing above and below, but very irregular in shape, being usually much more full on one side than the other. Skin: pale yellow, with an orange tint on the side next the sun, with distinct crimson spots irregularly placed, and with occasional cob-web streaks of russet. Eye: small and closed, with reflexed segments, very slightly depressed, and surrounded with five or more small distinct tubercles. Stalk: thin, half an inch long, set in a very narrow cavity, lined with thin pale russet. Flesh: white, moderately firm. Juice: plentiful, pale amber, sweetish and subacid, with some astringency.

The chemical analysis of the juice of the *Hagloe Crab* (season 1882), by Mr. G. H. With, F.R.A.S., F.C.S., Trinity College, Dublin, gave the following results :—

Density of fresh juice	1·057
Ditto after 24 hours' exposure to air ...	1·057
100 parts of juice by weight, yielded of	
Sugar	10·700
Tannin, Mucilage, Salts, &c.	2·110
Water ...	87·190

The *Hagloe Crab* seems to have disappeared from Herefordshire, for it has never once been shown at any of the apple shows of the last five years. It is still to be found in the parishes of Minsterworth, Westbury-on-Severn, Longney and Elmore, in Gloucestershire.

It will be seen on reference to page 26 of this work, that Mr. William Viner Ellis, of Minsterworth, has sent apples and grafts of this valuable variety to the Woolhope Club, and that Messrs. Cranston and Co., of King's Acre, near Hereford, have succeeded in propagating it.

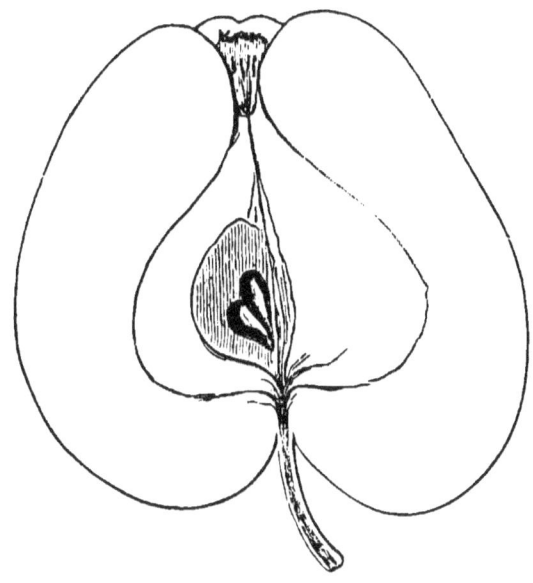

HANDSOME HEREFORD.

[SYN: *Handsome Norman; Bell Norman; La belle Normande.*]

This very distinct variety is not known in the Norman Orchards, although it seems to bear a French synonym. It is probably a Herefordshire seedling. Its real history is unknown.

Description.—Fruit: conical, snouted towards the apex, very uneven and irregular in its outline, being angular, and having especially one very prominent rib, which makes the fruit one sided; the base is rounded and swollen, so that the stalk is placed on an elevation of the surface. Skin: smooth, bright red on the side exposed to the sun, gradually fading towards the shaded side, where it is of a fine, deep, rich yellow; the whole surface is strewed with large russety specks, and the base surrounding the stalk, is covered with a patch of grey russet. Eye: closed, with erect pointed segments, set in a deep, irregular ribbed basin. Tube, long and conical; stamens, marginal. Stalk: short, sometimes half an inch long, inserted in a small, narrow cavity, Flesh:

yellowish, spongy, and sweet with astringency. Cells of the core, open; cell walls, elliptical.

The chemical analysis of the juice of the *Handsome Hereford* (season 1878), by Mr. G. H. With, F.R.A.S., F.C.S., Trinity College, Dublin, gave the following results:—

Density of fresh juice ...	1·051
Ditto after 24 hours' exposure to air	1·052
100 parts of juice by weight, yielded of	
Sugar	11·905
Tannin, Mucilage, Salts, &c.	4·038
Water	14.057

The *Handsome Hereford* makes a rich, deep coloured cider, which turns dark colour on exposure to air.

The tree is not large, but grows freely. It blossoms the middle of May, and ripens its fruit the end of October. It bears abundantly, and is grown extensively of late years.

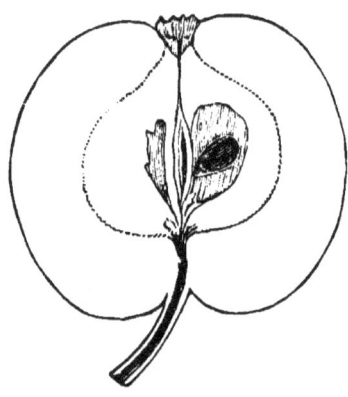

JOEBY CRAB.

[SYN: *Joby Crab.*]

A very old variety in Herefordshire, but without any known history. Its name is supposed to be a corruption of "jovial," a tribute to the strength of the cider made from it. When a labourer becomes merry from too much cider, its a rural pleasantry to say to him, "Ah! you've been in the sun, you be soon got joby."

Description.—Fruit: small and round, evenly shaped. Skin: almost entirely covered with deep bright crimson, except where shaded, and then it is deep greenish yellow, with a few stains of pale crimson and broken streaks of the same colour towards the exposed side. Eye: very small and closed, set in a shallow, plaited basin. Stalk: sometimes a mere knob, and sometimes slender, a quarter of an inch long, and rather deeply inserted. Flesh: white and firm. Juice: plentiful and thin, of a pale, pink colour, and a very acid and astringent taste.

The chemical analysis of the juice of the *Joeby Crab* (season 1881), by Mr. G. H. With, F.R.A.S, F.C.S., Trinity College, Dublin, gave the following results:—

Density of fresh juice ...	1·050
Ditto after 24 hours' exposure to air	1·055
100 parts of juice by weight, yielded of	
Sugar	10·300
Tannin, Mucilage, Salts, &c.	4·411
Water ...	85·289

The *Joeby Crab* is very highly esteemed in many orchards in Herefordshire, and this analysis proves its value, by showing the very large proportion of Tannin, Mucilage, and Salts which it contains. It is a very late fruit, and is scarcely fit for use before Candlemas. It makes a very strong cider, which it is often difficult to get bright. Being so late a fruit, it is frequently made alone for home use on the farm. It is used however more frequently to mix with other late apples to give the cider better keeping qualities; and it is added to late pears to give flavour and strength to the perry.

The *Joeby Crab* is to be found in most large orchards in Herefordshire, but the trees are usually old and cankered. Of late years it has not been much cultivated, though some of its admirers continue to propagate it.

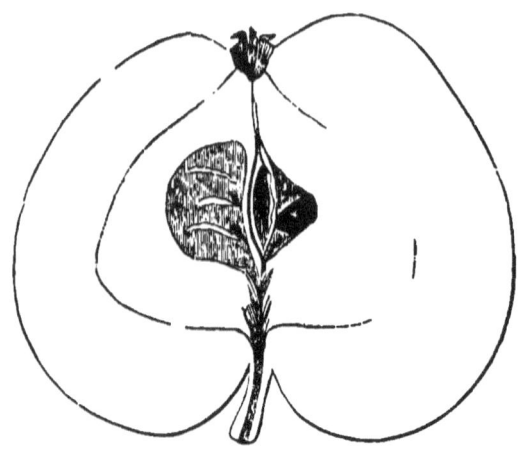

KINGSTON BLACK.

[Syn: *Black Kingston; Taynton Black; Taunton Black.*]

This valuable variety is believed to be a Somersetshire apple, and is said by tradition to have been raised at Kingstone, near Taunton. There is, however, no authentic record of its origin.

Description.—Fruit: of medium size, somewhat irregular in shape, two and a quarter inches broad, by two inches high, forming a short cone, broad and flat at the base; obscurely angular, and generally higher on one side of the apex than the other. Skin: of a dark mahogany or deep crimson colour, which extends over nearly the whole surface; where the colour is paler, it is splashed with broken streaks of dark crimson, and where shaded from the sun, the ground colour is deep yellow, approaching orange, and this is also marked with crimson streaks; the whole surface is strewed with fine cinnamon russet dots, and the base is generally covered with ashy grey russet, which often runs in streaks up the sides of the fruit. Eye: rather small, with erect segments, which are reflexed at the tips; stamens median; tube funnel shaped. Stalk: about a quarter of an inch long, inserted in a deep, russety cavity. Flesh: yellowish, with a pink tint near the skin, and fine-grained. Juice: plentiful, of a rich tawny red colour, with an agreeable aromatic flavour. It is moderately sweet, and pleasantly acid, with a strong astringent after-taste.

The chemical analysis of the juice of the *Kingston Black* (season 1881), by Mr. G. H. With, F.R.A.S., F.C.S., Trinity College, Dublin, gave the following results:—

Density of fresh juice	1·052
Ditto after 24 hours' exposure to air	1·055
100 parts of juice by weight, yielded of	
Sugar	10·028
Tannin, Mucilage, Salts, &c.,	6·792
Water	83·180

This valuable apple was introduced into Herefordshire by the late Mr. Palmer, who owned and occupied the estate at Bollitree, in the parish of Weston-under-Penyard, near Ross. Mr. Thomas Reynold, a nurseryman at Ross, procured the grafts for him from Somersetshire (c. 1820). Mr. Palmer planted an orchard of several acres with it, and so highly did he value the fruit, that he retained the orchard when the rest of the estate was sold. During the last 30 years, the present Mr. George Palmer, of Brooms Ash, has made large quantities of cider from the fruit, and has taken several first prizes with it at Hereford, at Gloucester, and at the Bath and West of England Agricultural Exhibition. In successful years he has sold it from the cask at 3s. the gallon, and at £1 1s. per dozen in bottle. In all good seasons it is worth 1s. 6d. per gallon, or £7 10s. the hogshead, and on one occasion a cask is reported to have been sold at £30, for bottling. Its general price has been from 9d. to 1s. 6d. a gallon in cask, and 12s. per dozen in bottles.

The tree grows to a middle size, and is spreading in character. It blossoms late, about the beginning of June, but is nevertheless rather a shy bearer. Its fruit is fit for gathering by the end of October, but does not become mellow and fit for the mill, until the first or second week in December. It is a very valuable variety, and its cultivation is extending very much throughout the county.

KNOTTED KERNEL.

This variety seems to have taken its name from the small knobs, or projections, round the eye.

Description.—Fruit : below medium size, round, but obscurely ribbed above, and having several small projections or knobs, round the eye. Skin: red throughout, getting much deeper and almost purple in colour on the side next the sun, and scattered with small distant specks of russet. Eye: closed, irregular in shape, slightly depressed. Stalk: slender, half an inch long, and nearly concealed in a deep narrow cavity lined with russet. Flesh: firm, slightly coloured beneath the skin, and along the core fibres. Juice: of a rich pink colour, becoming of a deep ruddy brown on standing, sweet, sub-acid and slightly astringent.

The chemical analysis of the juice of the *Knotted Kernel* (season 1882), by Mr. G. H. With, F.R.A.S., F.C.S., Trinity College, Dublin, gave the following results :—

Density of fresh juice ..	1·047
Ditto after 24 hours' exposure to air	1·051
100 parts of juice by weight, yielded of	
Sugar	11·700
Tannin, Mucilage, Salts, &c.	3·200
Water	85·100

This analysis shews that it is a valuable apple.

The tree is very hardy and bears well, and the deep colour of the fruit also makes it popular.

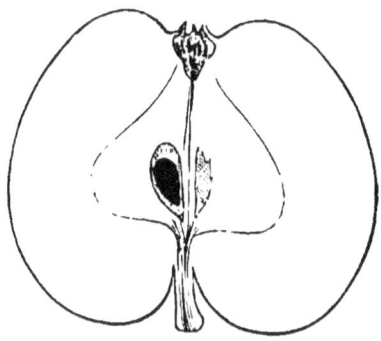

MÉDAILLE D'OR.

A seedling raised by Monsieur Goddard, of Boisguillaume, Rouen. A Gold Medal was awarded to its fruit in 1873 for its superior properties by the SOCIÉTÉ CENTRALE D'HORTICULTURE DU DÉPARTEMENT DE LA SEINE INFÉRIEURE. It was introduced into Herefordshire, in 1884, by the Woolhope Naturalists' Field Club.

Description.—Fruit: small, oblate, broad at the base, often irregularly spheroidal. Skin: golden yellow, almost completely covered with a marble work of thin brown russet, which often concentrates in patches, and becomes continuous round the eyes; there is often a slight touch of rose colour on the side next the sun. Eye: large and closed, sunk in a deep cavity, with slightly grooved borders. Stalk: thin and woody, about half an inch long, and inserted in a deep depression. Flesh: yellowish and tender. Juice: very sweet, with a strong, rough, astringent flavour, and not unpleasant.

The tree is very fertile and bears its fruit in clusters. In general appearance, and lightness of structure, this fruit resembles the old English variety *Forest Styre*. As a vintage fruit it takes the very highest rank. The juice attains the very high density of 1·102; and each kilogramme contains 238 grammes of sugar, giving 14 to 15 per cent. of alcohol; 5·509 of Tannin; and 1·428 of acid as compared with monhydrous sulphuric acid.

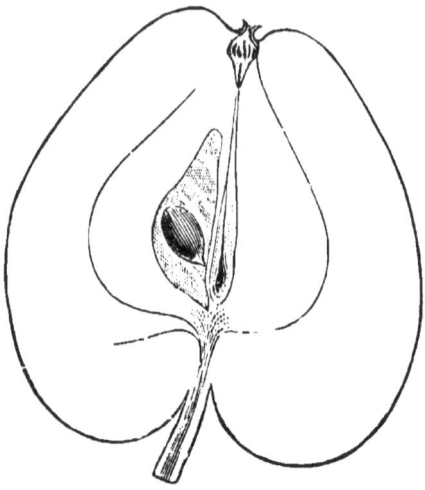

MICHELIN.

A seedling raised by Monsieur Legrand, Yvetot. It first bore fruit in 1872, and was dedicated by him to Monsieur Michelin, of Paris, Chevalier de la Légion d'Honneur, Member of the SOCIÉTÉ, CENTRALE D'HORTICULTURE DE FRANCE, ET DE LA SEINE-INFÉRIEURE, and one of the original promoters of the Congress appointed by the French Government for the study of Cider Fruits, and who attended all its meetings. This variety was introduced into Herefordshire, in 1884, by the Woolhope Naturalists' Field Club.

Description.—Fruit: of middle size, conical, with obtuse angles, becoming more marked as the fruit becomes more narrow towards the eye. Skin: green throughout, becoming yellowish green as it ripens; it presents a slight blush of red on the sunny side, and numerous small specks over the surface, with here and there a streak of russet. Eye: small and closed, almost level with the surface, and surrounded by a patch of light grey russet. Stalk: half an inch long, and inserted in a shallow cavity lined with russet, which spreads in streaks over the base of the apple. Flesh: white, tender, sweet, and rich.

" This is an apple of the highest merit," says Monsieur

Hauchecorne, "and is well worthy of extensive cultivation." The juice has a high colour and a density of 1·083. In 1·000 parts there are of alcoholic sugar 194; tannin 5·509; mucilage 11; acidity 1·071; salts, &c., 8·420; and water 780.

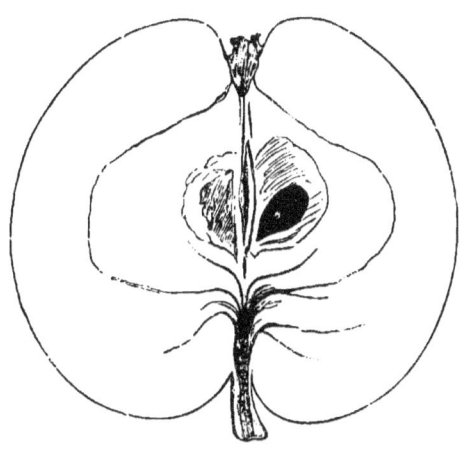

MUNN'S RED.

[SYN: *Pretty Maid; Greasy Apple.*]

This apple derives its name from that of its producer, a householder at Canon Pyon. It is widely grown in Herefordshire, and attracts attention in most orchards by the remarkably bright and glossy colour of its fruit.

Description.—Fruit: round, sometimes slightly ovate, even and regular in its outline. Skin: bright red, approaching scarlet, mottled, and somewhat streaked with crimson over the whole surface. Eye: closed, with convergent segments, set in a rather deep basin, which is sometimes even and saucer-like, and sometimes a little angular; tube, short, funnel shaped; stamens median. Stalk: long, curved, and rather stout and woody, inserted in a very deep, round cavity. Flesh: yellowish, with a stain of red from the base of the eye round the carpels. Cells of the core, open; cell-walls, elliptical.

The chemical analysis of the juice of the *Munn's Red* (season

1878), by Mr. G. H. With, F.R.A.S., F.C.S., Trinity College, Dublin, gave the following results:—

Density of fresh juice	1·0450
Ditto after 24 hours' exposure to air	1·0456

100 parts of juice by weight, yielded of

Sugar	9·110
Tannin, Mucilage, Salts, &c.	4·718
Water	86·178

Notwithstanding this analysis, its cider is not deemed of first excellence.

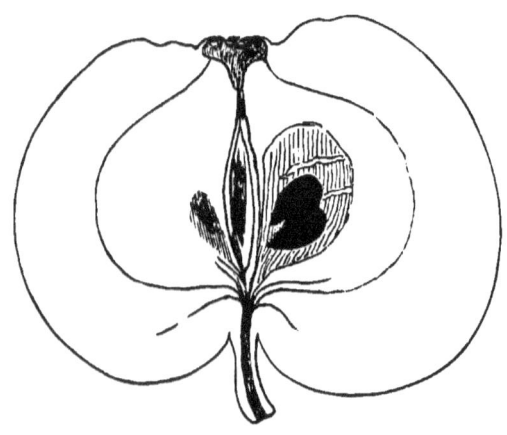

PYM SQUARE.

[SYN: *Izard's Kernel; Eggleton Red.*]

This variety originated at Eastnor Farm, near Eastnor Castle, Ledbury. Mr. Henry Izard some forty years ago (c. 1839), when staying there as a boy, planted three pips of an apple he was eating, in a flower pot. The seedlings were afterwards planted by Charles Bourne, the gardener from Ledbury, in a waste corner of the garden. In due course they were grafted on young crab stocks. This plant grew very vigorously, and bore fruit the second year after grafting. The two others proved worthless. Bourne called it *Izard's Kernel*, but it afterwards got the name of *Pym Square*, under the mistaken idea that it was a Devonshire apple introduced into Herefordshire. The origin of the name of *Pym Square* is not known. It is peculiar, and since there does not seem to be any Devonshire apple of that name, it is retained.

Description.—Fruit : rather over medium size, round, inclining to oblate, even and regular in its outline. Skin : smooth and shining, entirely covered with bright crimson, which is rather paler on the shaded side, and slightly mottled with yellow where the ground colour is visible. Eye : small and closed, with flat segments, and surrounded with small bosses, or knobs on the margin of the depression. Tube, funnel shaped ; stamens, marginal. Stalk : sometimes a mere knob, on the rounded base of the fruit ; at others half an inch long, slender, and inserted in a deep narrow cavity. Flesh : yellowish, tinged with red under the skin, very tender and juicy, briskly and well flavoured. Cells of the core, open; cell walls, obovate.

The chemical analysis of the of juice of the *Pym Square* (season 1878), by Mr. G. H. With, F.R.A.S., F.C.S., Trinity College, Dublin, gave the following results :—

Density of fresh juice ...	1·031
Ditto after 24 hours' exposure to air	1·035
100 parts of juice by weight, yielded of	
Sugar	10·219
Tannin, Mucilage, Salts, &c.	2·499
Water	87·282

The *Pym Square* apple has spread from Eastnor into the neighbouring orchards of Herefordshire and Worcestershire. Its fruit makes excellent cider, which has sometimes made the voyage to India with great credit to itself. The apples are brilliant in colour and good in flavour, so that they will sell to advantage as Table Fruit. "For culinary purposes they are excellent" says Mr. Izard "and as soon as the cook finds out their virtues, they are apt to prove bad keepers."

The tree is hardy and grows strong in the wood. It blossoms the beginning of May, and bears a good crop of fruit to ripen the middle of October.

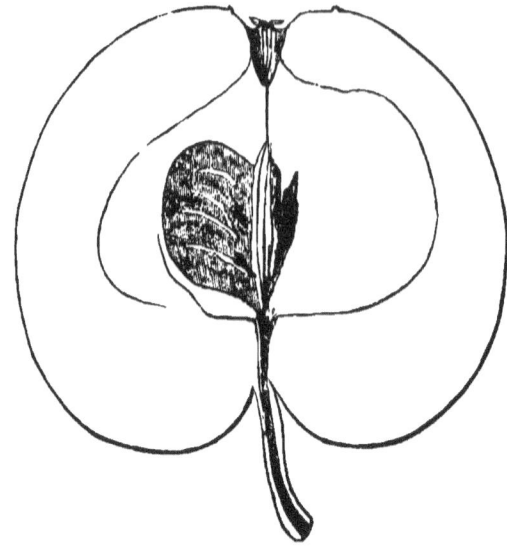

RED BUD.

Red Bud is a favourite name in the Herefordshire orchards, for red coloured, and otherwise unnamed fruit. At every apple show, three and sometimes four distinct apples have been shown under this name. There is nothing known definitely of the origin of any of them. They are probably seedlings from the orchard nursery, and propagated from their colour, and from their bearing virtues.

Description.—Fruit: medium size, roundish oblate, with obtuse angles, extending nearly to the base. Skin: smooth, bright red, much deeper on the side next the sun. It is however, puckered in lines about the eye, and they sometimes run down the ribs to the base. Eye: closed, with reflex segments, slightly depressed, in a basin puckered with folds of the skin, as well as slight fleshy tubercles. Stalk: thin, an inch long, set in a deep and narrow cavity, lined with russet. Flesh: yellow, tinged with red for some distance from the skin, soft, with a slightly acidulated taste. Juice: of full amber colour, viscid, and not abundant, sweet, with slight astringency.

The chemical analysis of the juice of the *Red Bud* (season

1883), by Mr. G. H. With, F.R.A.S., F.C.S., Trinity College, Dublin, gave the following results :—

Density of fresh juice	1·058
Ditto after 24 hours' exposure to air	1·060
100 parts of juice by weight, yielded of	
Sugar	11·120
Tannin, Mucilage, Salts, &c.	2·080
Water	86·800

The tree is hardy, grows to the full medium size, and bears profusely

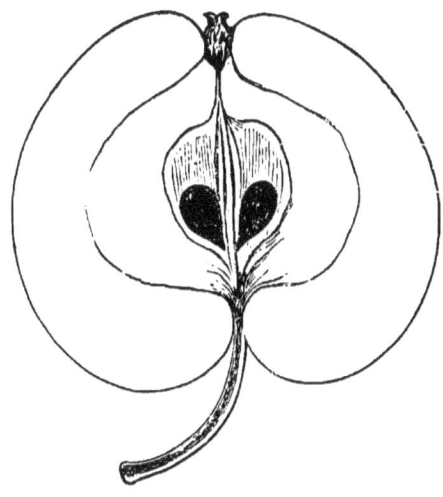

RED FOXWHELP.

This apple is chiefly grown in the Bodenham and Marden districts. It is pretty, well shaped, and very rich in colour. It is pleasant to eat, useful in cooking, and its growers value it as a cider apple.

Description.—Fruit : small, roundish ovate, even and regular in its outline. Skin : uniformly very dark crimson, almost of a chestnut or mahogany colour over its whole surface, except a small portion on the shaded side, which is a little, but very little paler. Eye : small and slightly open, with short, rather erect segments, and set in a shallow, plaited basin ; tube, short conical ; stamens, rather marginal. Flesh : yellow, deeply stained with crimson, both under the skin and at the core; very tender, pleasantly flavoured, and with a slight acidity. Cells of the core, open; cell walls, ovate.

The want of size in the *Red Foxwhelp*, and its want of sufficient character too, will prevent its being generally grown. Its chemical analysis, however, shows it to be rich in sugar and mucilage.

The chemical analysis of the juice of the *Red Foxwhelp* (season 1878), by Mr. G. H. With, F.R.A.S., F.C.S., Trinity College, Dublin, gave the following results:—

Density of fresh juice	1·043
Ditto after 24 hours' exposure to air	1·500
100 parts of juice by weight, yielded of	
Sugar	10·010
Tannin, Mucilage, Salts, &c.	4·256
Water	85·734

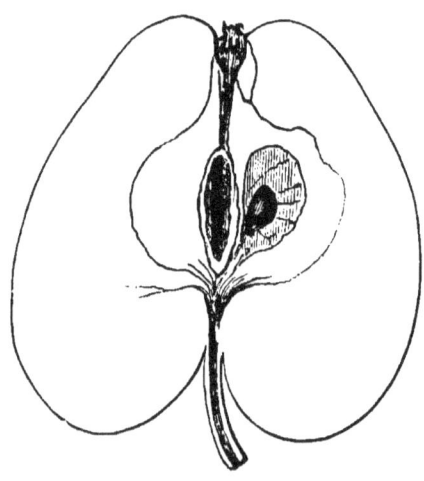

RED HEREFORD.

[Syn: *Red Norman.*]

This is an old variety in Herefordshire, and a very favourite one. Its history is not known.

Description.—Fruit: conical, or long conical, snouted and puckered towards the apex. Skin: smooth, lemon yellow, with a faint blush of red on the side exposed to the sun; the surface is sparingly strewed with minute russet points. Eye: very small, with convergent segments, set in a shallow and narrow puckered basin; tube very long and slender, funnel shaped; stamens, marginal. Stalk: half an inch long, slender, and obliquely inserted, frequently with a swelling on one side, at the base of the fruit. Flesh: greenish yellow, woolly, sweet, but not very juicy. Cells of the core, very large and closed; cell walls, ovate.

The chemical analysis of the juice of the *Red Hereford* (season 1878), by Mr. G. H. With, F.R.A.S., F.C.S., Trinity College, Dublin, gave the following results :—

Density of fresh juice	1·044
Ditto after 24 hours' exposure to air	1·051
100 parts of juice by weight, yielded of	
Sugar	11·905
Tannin, Mucilage, Salts, &c.	3·942
Water	84·153

The *Red Hereford* is held in much esteem in the Herefordshire Orchards, and is widely grown. It makes a rich cider, dark in colour, with a rich, sweet, and highly aromatic flavour.

The tree is vigorous and fertile, but not large. It blossoms the middle of May and bears freely, ripening at the end of October.

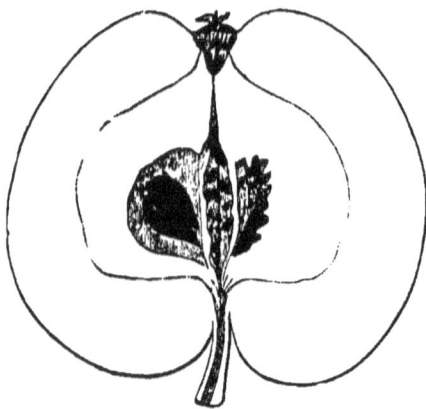

RED ROYAL.

A favourite apple in the Gloucestershire Orchards, without known history.

Description.—Fruit: small, roundish, inclining to oblate, and sometimes to ovate, bluntly angular on the sides. Skin: almost entirely covered with dark crimson, except on the shaded side, where it is yellow; the surface is sprinkled with russety dots. Eye: quite closed, with convergent segments; tube, funnel shaped; stamens, median. Stalk: short and slender, inserted in a rather deep cavity. Flesh: white and tender. Juice: plentiful, pale in colour, sweet, but slightly bitter, and pleasantly subacid. Cells of the core, open.

The chemical analysis of the juice of the *Red Royal* (season 1880), by Mr. G. H. With, F.R.A.S., F.C.S., Trinity College, Dublin, gave the following results:—

Density of fresh juice ...	1·035
Ditto after 24 hours' exposure to air	1·037
100 parts of juice by weight, yielded of	
Sugar	13·700
Tannin, Mucilage, Salts, &c. ...	·260
Water	86·040

This variety is highly esteemed in Gloucestershire. It is thought to make cider of the first quality, of good colour and flavour, and very sweet and pleasant. The analysis does not indicate so high a character, for though the juice abounds in sugar, it is very deficient in tannin, mucilage and salts, to which good body, and keeping qualities are attributed. The fruit meets with a ready sale in the market both for edible and culinary purposes.

The tree is hardy, and bears well. It likes a high situation and deep strong loam, well drained, as indeed do most apples of character.

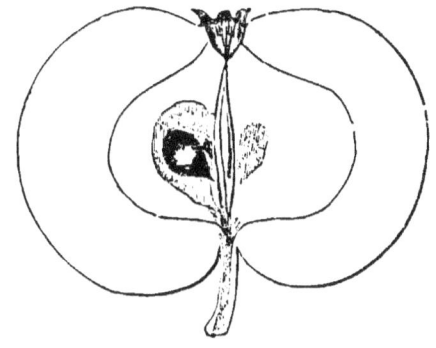

RED SPLASH.

[Syn: *Rolling's Kernel.*]

The origin of this pretty variety is nowhere given, though it is extensively cultivated in some Worcestershire Orchards.

Description.—Fruit: small, roundish oblate, and regularly formed. Skin: golden yellow, with a few streaks of crimson on the shaded side, and completely covered with crimson on the side exposed to the sun. Eye: with divergent segments reflexed at the top, set in a wide and saucer-like basin. Tube, short, funnel shaped; stamens, median. Stalk: a quarter to half an inch long, slender, and set in a rather wide cavity. Flesh: yellowish, juicy, sweet, and agreeably flavoured, but with considerable astringency. Cells, open; cell walls, roundish, inclining to obovate.

This small apple is a valuable early cider fruit.

The chemical analysis of the juice of the *Red Splash* (season

1881), by Mr. G. H. With, F.R.A.S., F.C.S., Trinity College, Dublin, gave the following results:—

Density of fresh juice	1·042
Ditto after 24 hours' exposure to air	1·043
100 parts of juice by weight, yielded of	
Sugar	9·600
Tannin, Mucilage, Salts, &c.	4·790
Water	85·610

The tree makes a sturdy standard of small size. It is very hardy and prolific. It is grown very much in the neighbourhood of the Malvern Hills; and the fruit from the parish of Newland finds a ready sale, it is said, for the manufacture of "Real Chutnee Sauce."

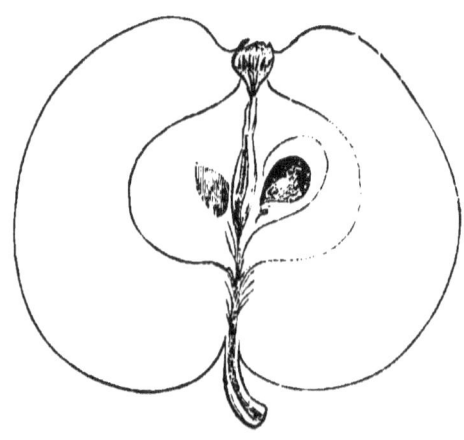

REDSTREAK.

[SYN : *Scudamore's Crab ; Herefordshire Redstreak ; Redstrake of King's Caple ; Irchinfield Redstrake.*]

"Yours be the produce of the soil :
O ! may it still reward your toil !
But though the various harvest gild your plains,
  Does the mere landscape feast your eye ?
Or the warm hope of distant gains

> Far other cause of glee supply?
> Is not the Redstreak's future juice
> The source of your delight profound
> Where Ariconium pours her gems profuse
> Purpling a whole horizon round."
>
> SHENSTONE.

The *Redstreak* has been the most fortunate of all cider apples for the renown it has obtained. It appeared at a time when the greatest attention was paid to the Herefordshire Orchards. It at once found a patron of remarkable energy and influence, and its praises have been said and sung, in prose and verse, beyond all other apples. It seems to have originated about the beginning of the 17th century, and was first brought into general notice by Lord Scudamore. Evelyn (1706) is the first author who mentions it as "the famous *Red-strake* of Herefordshire, a pure Wilding within the memory of some now living, surnamed the *Scudamore's Crab*, and then not much known, save in the neighbourhood. Phillips next took up its praise, and in his poem *Cyder* says:—

> " Let every tree in every garden own
> The Redstreak as supream; whose pulpous fruit
> With Gold irradiate, and Vermillian shines.
> Tempting, not fatal as the birth of that
> Primæval interdicted Plant, that won
> Fond *Eve* in hapless hour to taste and die.
> This of more bounteous Influence inspires
> Poetic Raptures, and the lowly Muse
> Kindles to loftier strains; even I perceive
> Her sacred Virtue. See! the numbers flow
> Easie whilst cheer'd with her Nectarious Juice,
> Her's and my Country's Praises I exhalt.
> Hail Herefordian Plant! that does disdain
> All other fields! Heaven's sweetest Blessings, hail!
> Be thou the copious Matter of my Song,
> And thy choice Nectar; on which always waits
> Laughter and Sport, and care beguiling Wit,
> And Friendship, chief Delight of Human Life,
> What should we wish for more."

The *Redstreak* apple was thus brought into the highest popular favour, and its sweet and pleasant cider was deemed "a fitting present for Princes." It completely supplanted the *Gennet Moyle*, which had before held the palm in public favour, and indeed for the time being, all other cider apples were thrown into the shade. How widely it was cultivated is well shewn from the following extract from a M.S. in the Bodleian Library, at Oxford, entitled

> " *The History of Gloster, or the Antiquities, Memoirs, and Annals of ye ancient City and Royal Dukedom of Gloster from its original to the present time* "
>
> by ABEL WANTNER, Citizen of Gloster, 1714:
>
> "Dimock and Kemply, before mentioned, are two of the most note'edst parishes in England for making the most and best rare *Vinum Dimocuum*, or that transcendant Liquor, called Redstrake Sider, not much inferior to the best French wines. And so plentiful that old Master Wyniat, of the Grainge, (a worthy gentleman and a noble housekeeper,) hath caused but one apple to be gathered from each Apple Tree growing in his own Grounds, and with the Liquor thereof he hath made a whole hogshead reare good Sider."
>
> *Furley MS., Vol. iv., fol.* 196, *p.* 2.

This extract appears in the parish Register of Dymock, but without the reference and date here given.

The reputation of the *Redstreak* apple began to decline about the middle of the 18th Century. Its cider, though sweet and pleasant, had not much strength, and would not keep well. "Its Liquor," Nourse describes, as "of noble colour and smell, but withal very luscious and fulsome. They who drink it will find their stomachs pall'd sooner by it, than warmed and enliven'd." In justice to the Herefordshire cider makers, it must however be stated, that its cider was thought from the very first to be inferior in strength and quality, to that made from many other kinds of fruit. "*Gennet Moyle* makes the best cyder in my judgment, and such as I do prefer before the much commended *Redstreak'd*" says Dr. Beale (1656), and Evelyn, and the writers in the Appendix to his "*Pomona*," say as much for several other apples, as *Woodcock, Hagloe Crab, Underleaf, Styre, Must, Bromsborrow Crab, &c.* The soundness of this judgment was soon confirmed by experience, for by the end of the last century the *Redstreak* had quite lost all favour. Dunster in his "*Notes to Phillips Poems*," thought the true method of managing was lost, for out of ten or twelve casks seldom more than two or three proved good, and adds " it is now (1791) seldom made from." Marshall (1796) says plainly "the *Redstreak* apple is given up, and Thomas Andrew Knight speaks of it (1811) as having survived its good qualities.

Description.—Fruit : middle sized, two inches and three quarters wide, and two inches and a quarter high ; roundish, narrowing towards the apex. Skin : deep, clear yellow, streaked with red on the shaded side, but red, streaked with deeper red, on the side next the sun. Eye : small, with convergent segments, set in a rather deep basin. Stalk : short and slender. Flesh : yellow, firm, crisp, and rather dry. Specific gravity, or density of the juice 1·079 (Knight).

The tree seems naturally to have been very short lived. It was low, shrubby and rugged in growth. Evelyn says of it "That as the best Vines of richest liquor and greatest burthen do not spend much in Wood and unprofitable branches, so nor does this tree."

The result of careful enquiries recently made for the true old variety, was the discovery of only one tree at King's Caple, which however was blown down in the spring of 1878. Mr. Reginald Wynniat, of the Grainge, Dymock, has kindly ascertained (1879) for this paper, that there is still one tree remaining at Kempley, of the many thousands growing at Dymock and Kempley, in 1714. *Redstreak's* there are in abundance in every parish, named simply from their mode of colouration. Evelyn noticed this fact "The *Red-strake* of King's Caple, and those parts, is of great variety, some make cider that is not of continuance, yet pleasant and good ; others that lasts long, inclining towards the *Bromsborrow Crab* rather than a *Red-Strake.*"

The old *Redstreak* as a distinct variety has now ceased to exist, and it may be added that its loss is not to be lamented.

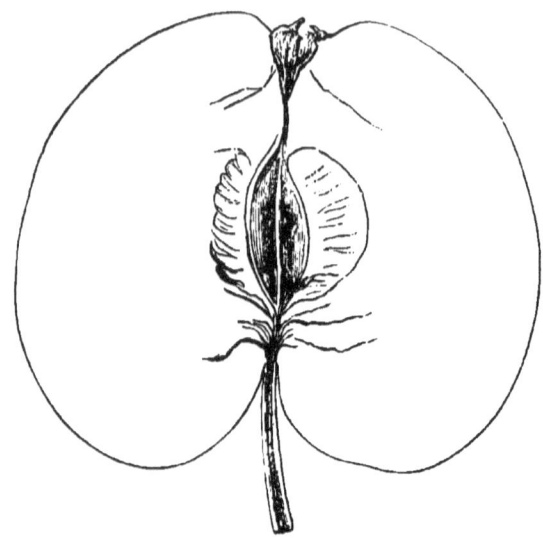

REJUVENATED FOXWHELP.

[SYN: *The Canon Apple; Crow's Kernel.*]

This apple is one of peculiar interest in Herefordshire. The epithet "new" will be used in treating of it, but merely to prevent any confusion in discussing the question, as to whether it is "new," that is, a seedling; or whether it is the true "old" *Foxwhelp*, restored to a flourishing rejuvenated form by a careful system of grafting and re-grafting. This is a question that has been warmly discussed by the growers for some years past.

At first sight, the distinction between them seems very marked; both the tree of the "new" *Foxwhelp* and its apple, are much more luxuriant than the "old" *Foxwhelp*. The apple of the "new" *Foxwhelp* is not only larger, but in its general character it is broad in shape—or in other words, its lateral is greater than its longitudinal diameter—whilst the apple of the "old" *Foxwhelp* is smaller, and usually oblong in shape ; but on a careful examination of the trees of either kind, the apples are so similar in shape and appearance, that it would be impossible to distinguish them, if thrown together. The difference of size and shape is due simply to the improved vitality and luxuriance of the growth of the tree. The points of similarity between them are very striking. There is the same brilliant colour ; the same tough, leather-like skin ; the same eye ; the same long slender stalk set in its deep, narrow channel ; and to this it may be added, they have the same period of arriving at maturity. Then again, the chemical analysis shows no greater difference between them, than may be accounted for, by the more watery juice of the fruit of the more free growing tree.

The history of the "new" *Foxwhelp* can be traced with some clearness. A farmer of the name of Yeomans, living at Cowarne, between 60 and 70 years ago, took an unusual interest in the "old" *Foxwhelp*, and both in that parish, and at Canon Pyon, to which he afterwards migrated, he grafted and re-grafted it on healthy stocks, until he restored its luxuriance of growth. Another farmer, a Mr. Crowe, and Messrs. Skidmore, Miles, and Williams, wheelwrights, of Canon Pyon, systematically, but separately, carried on the system of re-grafting, beginning on seedlings of the "old" *Foxwhelp*. Their success had been well established by 1823, when first Mr. Jay, of Lyde, and afterwards Mr. Bosley, of Lyde, and Mr. Hill, of Eggleton, obtained grafts, and thus from the centres of Cowarne and Canon Pyon, intelligent fruit growers got their supply of grafts, and we have the handsome, luxuriant, and useful fruit of this time. With this distinct history there can scarcely be a doubt, that the "new" *Foxwhelp* is simply the "old" historic variety rejuvenated by careful management, but the disbelief in this has arisen from the absence, in part or altogether, of the true *Foxwhelp* flavour in the cider made from it, which is so remarkable and characteristic in the "old" *Foxwhelp*. As a matter of fact, its cider is more sweet and luscious than that made from the "old" *Foxwhelp*, and in flavour resembling far more the cider made from the *Cowarne Red* apple.

It must be remembered, however, that sometimes for years together, the cider from the "old" *Foxwhelp* itself gives but a faint suspicion of the true *Foxwhelp* flavour which is so highly esteemed, and moreover, that it is only of late years, comparatively speaking,— that is, after the trees had become of considerable age—that the cider has gained the pride of place it now so deservedly holds. In Evelyn's time, the "old" *Foxwhelp* was merely considered a first-class cider fruit. It must be left therefore to time, to develope the true flavour of the *Foxwhelp* in its rejuvenated form.

The chemical analysis of the juice of the *Rejuvenated Foxwhelp*, by Mr. G. H. With, F.R.A.S., F.C.S., Trinity College, Dublin, gave the following results:—

Density of fresh juice	1·043
Ditto after 24 hours' exposure to air	1·044
100 parts of juice by weight, yielded of	
Sugar	8·000
Tannin, Mucilage, Salts, &c.	4·301
Water	87·699

The *Rejuvenated Foxwhelp* has intrinsic merits of its own, and for this cause alone, it should be grown much more plentifully than it has hitherto been. Every orchard should possess it, and its owners may await with good faith, the development of the true *Foxwhelp* flavour in its cider, as the trees grow older. Speaking of it as an apple, it may be said, that it is above the medium size, and its brilliant colour recommends it to every one. It sells well in September as a "pot fruit." It has a piquant, acid, rough flavour, which would not please all palates to eat raw, but as a cooking apple, it is excellent for pies and puddings; and "the apple of all others to make sauce for the Michaelmas goose, or for a roast leg of pork."

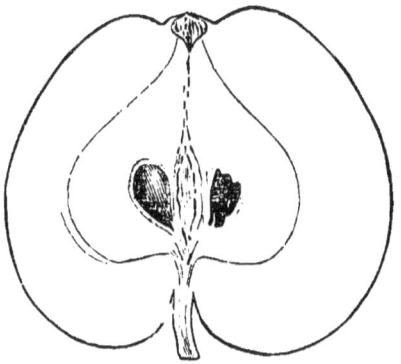

ROUGE BRUYÈRE.

An old variety, whose history is not known. Its name is often given to other Apples (varieties of *Argile* or *Fréquin*, &c.,) a fact which proves the general esteem in which it has long been held. It was introduced into Herefordshire in 1884, by the Woolhope Naturalists' Field Club.

Description.—Fruit: small and symmetrical, broad at the base, becoming slightly angular at the upper third. Skin: almost entirely carmine, deeper on the sunny side, and having small grey spots scattered over the surface. Eye: small and closed, set in a very shallow depression. Stalk: short and woody, inserted in a narrow cavity, lined with russet, which also spreads over the base of the apple. Flesh: whitish yellow, firm, with a sweet juice, a bitter, pleasant taste, and an excellent aroma.

This is a very favourite apple throughout the orchards of Normandy. "It is superior," says Monsieur Hauchecorne, "to all others bearing its name, and makes excellent cider without mixture with other fruits." Its esteem is only equalled by the *Argile Grise*. The abundance of tannin in the juice, renders it very valuable to give good keeping qualities to the cider from mixed fruits. The density of the juice is 1·075 to 1·080. In 1·000 parts it contains of alcoholic sugar 175; tannin 7; mucilage 8; acidity as compared with monhydrous sulphuric acid 1; salts, &c., 9; and water 800.

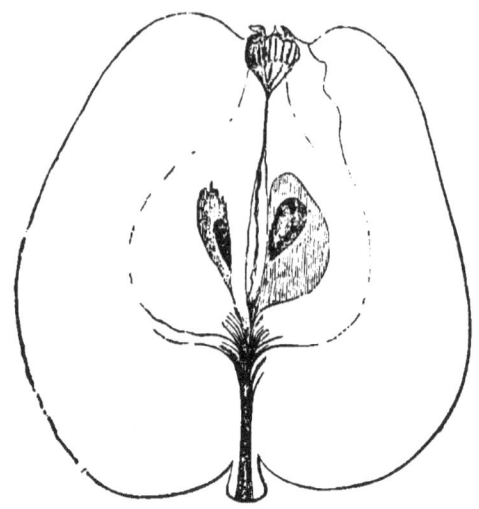

ROYAL WILDING.

[SYN: *The Cadbury.*]

There is no published account of the origin of this variety. It is not mentioned by any of the cider authorities of the last century, and nothing is known of its history. There are, however, many large and old trees scattered throughout the county, which proves that it must have been in existence earlier than the present century. In Somersetshire this apple is called " *The Cadbury.*"

Description.—Fruit : of middle size, conical, contracted round the upper third ; with obtuse ribs on the sides which extend to the crown, and form ridges round the eye. Skin : greenish yellow on the shaded side, and brownish red on the side next the sun. The whole surface is often covered with very small specks of a green tint on the shaded side, and red where coloured. Eye : small, set in a narrow puckered basin, and with convergent segments, Stalk: short, or a mere knob deeply inserted in a shallow cavity, often lined with thin pale russet. Flesh : woolly and tough, not very juicy. The juice has a vapid, bittersweeet flavour, with but little acidity, and is very dark in colour.

The chemical analysis of the juice of the *Royal Wilding* (season 1880), by Mr. G. H. With, F.R.A.S., F.C.S., Trinity College, Dublin, gave the following results:—

Density of fresh juice ...	1.037
Ditto after 24 hours' exposure to air	1·039
100 parts of juice by weight, yielded of	
Sugar	10·712
Tannin, Mucilage, Salts, &c.	4·688
Water ...	84·600

The specific gravity of the fresh juice of this variety from fruit grown in 1876, was 1·066; in 1878, it was 1·056; and in 1881, 1047; all considerably higher than that for 1880, facts which show how great is the influence of sunshine, in the production of sugar.

The *Royal Wilding* is a late fruit, and holds a high place in general esteem as one of the most useful varieties. It is deficient in flavour by itself, but its value is derived from the body and strength it gives to the cider, when mixed with other varieties, whose juices supply a higher flavour.

The tree is hardy, very full of leaves, and forms a wide-spreading handsome head, but it is generally thought to be a shy bearer. "When the *Royal Wilding* bears well," says a Hereford-shire proverb, "it is always a good cider year;" meant not so much in compliment to the fruit itself, as to show that this variety requires a favourable season.

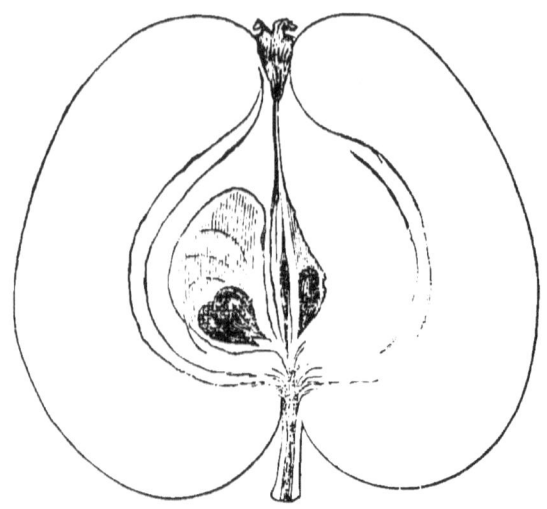

SACK APPLE.

[SYN: *Spice Apple; Fox's Kernel.*]

This apple is one of our oldest historic varieties.

Description.—Fruit, below medium size; conical, and uneven in its outline, being ribbed on the sides like the Margil, and rigid round the eye. Skin: smooth and shining, as it varnished, almost entirely covered with deep bright crimson, which is streaked and mottled with darker crimson on the side next the sun; but where shaded, it is yellowish and mottled with crimson. Eye: small and closed, with erect pointed segments, set in a deep and plaited basin; tube, funnel shaped; stamens, median; the style very stout and thick at the base, filling nearly the half of the tube. Stalk: very short, thick, and fleshy, set in a very shallow cavity. Flesh: tender, crisp, fine-grained, sweet, and with a pleasant, sub-acid flavour. Cells of the core, open; cell walls, ovate. In use during October and November.

The chemical analysis of the juice of the *Sack Apple* (season

1878), by Mr. G. H. With, F.R.A.S., F.C.S., Trinity College, Dublin, gave the following results:—

Density of fresh juice ...	1·036
Ditto after 24 hours' exposure to air	1·044
100 parts of juice by weight, yielded of	
Sugar	6·400
Tannin, Mucilage, Salts, &c.	5·220
Water	88·380

The *Sack Apple* is more useful in the present day as a dessert or pot fruit, than for cider. It is an early apple, but keeps fairly well. It has a pleasant, vinous, aromatic flavour.

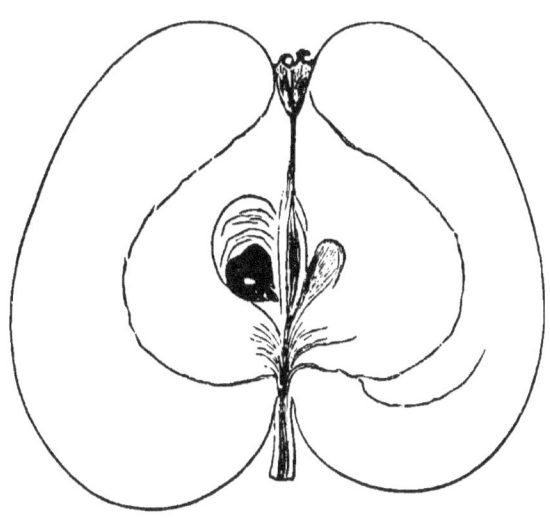

SAM'S CRAB.

[SYN: *Longville's Kernel.*]

This apple, according to Mr. Lindley, was originated in Herefordshire, where, he adds, curiously enough and very erroneously, "it is at present but little known." It is on the contrary, well known in Herefordshire, widely distributed, and very highly esteemed as a very early dessert fruit. It is used also for cooking, and for cider.

Description.—Fruit: conical, or roundish ovate, even and regular in its outline. Skin: beautifully streaked with crimson, and yellow on the side next the sun, and less so on the shaded side, where it is more yellow. Eye: closed, with connivent segments, set in a rather deep, round, and somewhat plaited basin; tube, funnel shaped; stamens, median. Stalk: about an inch long, slender, inserted in a deep cavity, which is tinged with green. Flesh: yellowish, tender, juicy, sweet, and of good flavour. It is tinged with red at the base of the eye, at the base of the stalk, and round the carpels. Cells of the core, open; cell-walls, ovate.

The chemical analysis of the juice of the *Sam's Crab* (season 1878), by Mr. G. H. With, F.R.A.S., F.C.S., Trinity College, Dublin, gave the following results:—

Density of fresh juice ...	1·037
Ditto after 24 hours' exposure to air	1·046
100 parts of juice by weight, yielded of	
Sugar	10·140
Tannin, Mucilage, Salts, &c.	4·370
Water	85·490

Sam's Crab is one of the most useful of all our useful apples. It requires a warm soil and sunny situation to bring its fruit to perfection. In unfavourable situations it could hardly be recognised as the same apple. When well ripened it has a rich aroma, and a juicy, sweet, and piquant flavour that is seldom equalled. It is a prime favourite with all Herefordshire school children (no mean judges of a good apple), and it is equally attractive to birds and insects, who revel in its sweetness. There are undoubtedly two varieties of this apple, or, as was quaintly expressed by a great admirer of the fruit, "There are two sorts of *Sam's Crab:* a basket full of one kind is eaten the same day, but the same basket full of the other kind lasts three or four days."

CIDER APPLES.

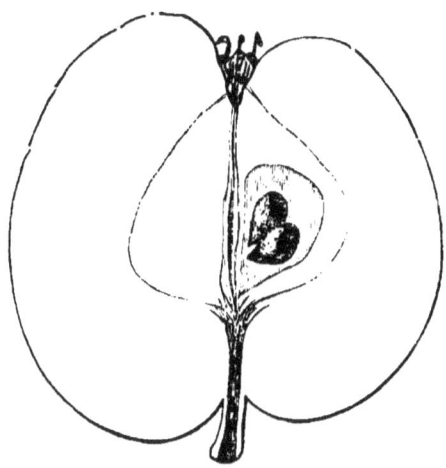

SKYRME'S KERNEL.

The Skyrmes are an old Herefordshire family, and a century or two since, one branch held an estate at Brockhampton, called the Upper House, for some generations. It passed to the Protheroes, by marriage, in 1788. Another branch of the Skyrmes lived at Dewsall, near Hereford. History is silent as to which of them grew the *Kernel* that bears the family name, but it may very probably have been raised at Brockhampton, for there are many trees there of some 100 or 150 years old; they are found in that district of the county, and may have spread from it. The apple is not mentioned by any of the old writers.

Description.—Fruit: small, about two inches wide, and two inches high, ovate, or slightly conical, even and regular in its outline, and sometimes snouted towards the apex. Skin: smooth and shining, almost entirely covered with broken streaks of brilliant crimson, on a thin pale crimson ground, on the side next the sun; and lemon yellow, tinged with crimson, and marked with pale crimson stripes, on the shaded side; the whole surface being strewed with distinct russet dots. Eye: small, set in a narrow, round, and even basin; segments, connivent; tube, funnel shaped; stamens, marginal. Stalk: short, on a fleshy knob, set in a deep wide cavity. Flesh: yellowish, firm, crisp, but not very juicy, with an acid, and rather harsh flavour; cells of the core, closed.

The chemical analysis of the juice of the *Skyrme's Kernel* (season 1880), by Mr. G. H. With, F.R.A.S., F.C.S., Trinity College, Dublin, gave the following results:—

Density of fresh juice ...	1·034
Ditto after 24 hours' exposure to air	1·037
100 parts of juice by weight, yielded of	
Sugar	10·638
Tannin, Mucilage, Salts, &c.	3·662
Water	85·700

This variety is very highly esteemed, and thought by some cider makers to be second only to the *Foxwhelp*, and to partake somewhat of its character. Its cider has a peculiar flavour, and its aroma improves very much by keeping; but it is better mixed with other apples of its season, such as the *Styre, Strawberry Hereford*, &c. When made by itself, the cider has the disadvantage (like the *Foxwhelp*) of turning dark on exposure to air, when in the glass on the table. *Skyrme's Kernel* is also sold, by its growers, as a culinary fruit, and gives a special flavour to pies and puddings, though it does not come into the market in this character.

The tree is hardy, and grows to a large size, with a wide-spreading growth. It blossoms at the end of May, is rather shy in bearing, and reaches maturity at the end of October, or beginning of November.

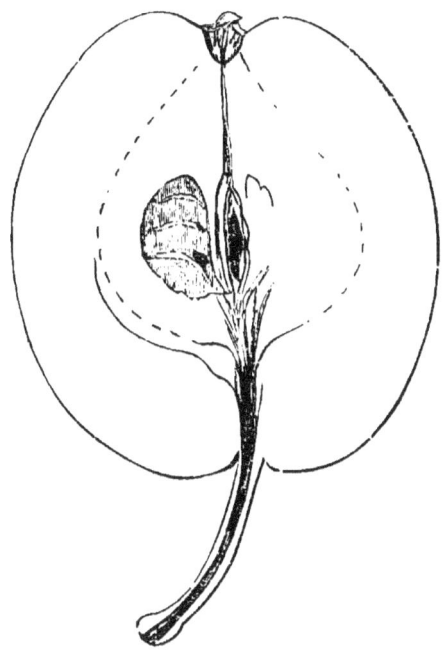

SOUTH QUEENING.

A favourite apple in the Herefordshire Orchards, but it is without any history. A "Queening," doubtless from its irregular and often angular shape, from "coin," or "coign," an angle.

Description.—Fruit: of medium size, roundish oblong, but of irregular shape. Skin: yellowish green, covered more or less by a blush of crimson, with streaks and marks of deeper colour. Eye: large, and closed, with thick, green, inverted segments, and seated in a narrow plaited basin. Stalk: three quarters of an inch long, inserted apparently on the surface, but really inclosed by the flesh of the apple. Flesh: white and soft, with a sweet acid taste, and some astringency. Juice: fairly plentiful, of a full amber colour, sweet and rather astringent.

The chemical analysis of the juice of the *South Queening* (season 1882), by Mr. G. H. With, F.R.A.S., F.C.S., Trinity College, Dublin, gave the following results:—

Density of fresh juice	1·050
Ditto after 24 hours' exposure to air	1·054
100 parts of juice by weight, yielded of	
Sugar	13·600
Tannin, Mucilage, Salts, &c.	1·733
Water	84·667

As a cider fruit it is very useful when mixed with varieties of rougher character and better keeping qualities. It is a grand fruit for cooking, and makes excellent sauce.

The tree grows well and in good form. It is very hardy, and a good bearer.

REDSTREAK APPLES.

The number and variety of *Redstreak Apples* is infinite. They abound in the orchards of Herefordshire, as they doubtless do elsewhere. The "Redstreak" is the most frequent form of coloration in the apple: and thus, when the seedling tree first bears fruit, or when an apple has no other recognised name, if the sun paints on it freely the bright streaks of crimson which are so attractive, it naturally takes the name of "Redstreak," with any other epithet, that may serve to distinguish it. Those "Redstreaks" only, which have obtained a character in the orchards from the virtue of their juices, will be noticed here. The renowned "Redstreak" of Lord Scudamore is no longer grown. It has long since been surpassed by superior varieties.

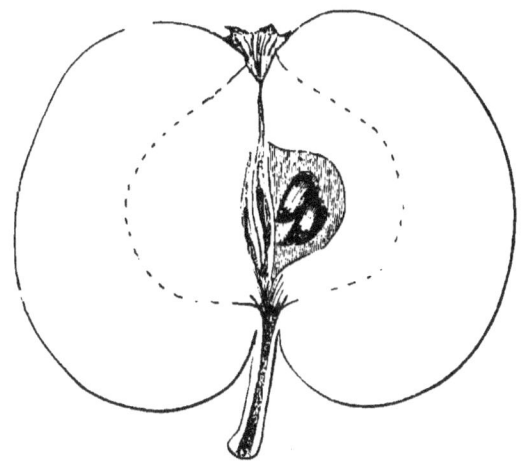

SPREADING REDSTREAK.

A variety so named from the spreading growth of the tree, and the colour of its fruit. It has no known history.

Description.—Fruit: full medium size, roundish oblate, and regular in shape. Skin: greenish yellow, with a deep blush next the sun, and streaks of darker crimson. Eye: open, in a wide, shallow cavity; calyx segments, short and reflexed; anthers projecting from the centre. Stalk: slender, half an inch long, inserted in a narrow and regular cavity. Flesh: soft and sweet, with a rough acidulated taste. Juice: very pale straw colour.

The chemical analysis of the juice of the *Spreading Redstreak* (season 1881), by Mr. G. H. With, F.R.A.S., F.C.S., Trinity College, Dublin, gave the following results:—

Density of fresh juice	1·049
Ditto after 24 hours' exposure to air	1·053
100 parts of juice by weight, yielded of	
Sugar	11·600
Tannin, Mucilage, Salts, &c.	3·400
Water	85·000

This analysis proves its value as a cider fruit, not only from the quantity of its sugar, but also from the amount of Tannin, Mucilage and Salts, which ensure its good qualities.

The tree grows to a large size, is very hardy, and bears well. It is much grown in the valley of the river Froome, where the trees are, many of them, nearly a century old.

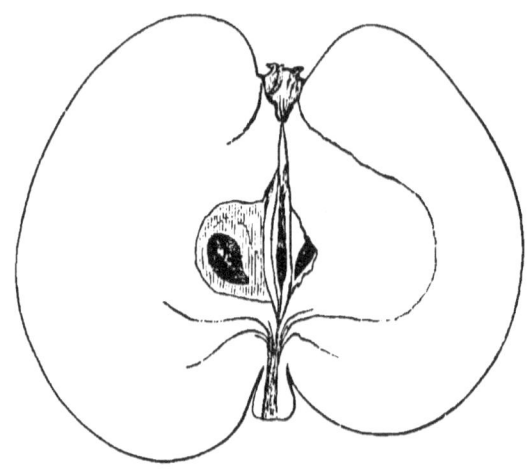

STRAWBERRY HEREFORD.

[SYN: *Strawberry Norman.*]

The origin of this apple is not known. It is probably a local Herefordshire seedling, and is now widely distributed throughout the county.

Description.—Fruit: small, round, flattened, and uneven in outline, being angular and considerably ribbed around the eye. Skin: with a lemon yellow ground, covered with light crimson, which is thickly marked with broken streaks and mottles of a bright and darker crimson, on the sunny side; and these streaks gradually getting more pale, are extended to the shaded side of the fruit; the stalk cavity and the base of the apple are lined with cinnamon-coloured russet. Eye: of middle size, with long, leafy, rather erect, and slightly divergent segments, set in a very deep and ribbed basin; tube, short and funnel shaped; stamens, inclining to basal. Stalk: very short, quite embedded in the cavity, which is lined with russet extending over the base. Flesh: yellowish, close and spongy, with a sweet mawkish juice. It has a crimson stain at the base of the eye. Cells of the core, small and closed; cell walls, obovate.

The fruit is pleasant in taste, and, when fresh, is supposed to have a slight suspicion of the flavour of the Strawberry.

The chemical analysis of the juice of the *Strawberry Hereford* (season 1878), by Mr. G. H. With, F.R.A.S., F.C.S., Trinity College, Dublin, gave the following results :—

Density of fresh juice	1·043
Ditto after 24 hours' exposure to air ...	1·045
100 parts of juice by weight, yielded of	
Sugar	13·736
Tannin, Mucilage, Salts, &c.	1·071
Water	85·193

The tree grows freely, blossoms the middle of May, and ripens its fruit the end of October. It bears well, and the fruit makes excellent cider. It deserves its popularity.

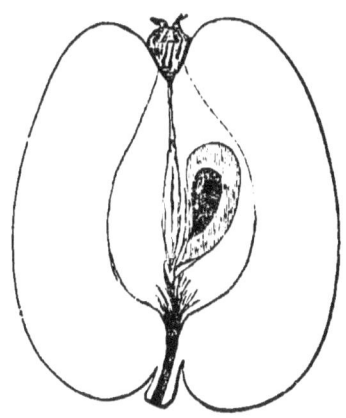

STYRE WILDING.

This fruit is without a history. It is widely grown, and many of the trees are more than a hundred years old.

Description.—Fruit: small, conical, bluntly angular, and irregular in its outline. Skin: smooth and shining, greenish yellow on the shaded side, and with a red cheek wherever exposed to the sun. Eye: closed, with connivent segments, set in a rather deep, narrow and plaited basin; tube, conical, sometimes inclining to funnel shape; stamens, median. Stalk: very short, deeply embedded in the cavity, which is russety, and generally with a fleshy swelling on one side of it. Flesh: soft and woolly, sweetish and scarcely acid. Cells of the core, open.

The chemical analysis of the juice of the *Styre Wilding* (season 1880), by Mr. G. H. With, F.R.A.S., F.C.S., Trinity College, Dublin, gave the following results:—

Density of fresh juice	1·041
Ditto after 24 hours' exposure to air	1·044
100 parts of juice by weight, yielded of	
Sugar	14·121
Tannin, Mucilage, Salts, &c.	·679
Water	85·200

This tree blossoms the end of May and ripens its fruit late. It is highly esteemed in some districts of the county, and is thought to give strength and flavour to the mixed fruit. With *Skyrme's Kernel* and the *Redstreak* it makes a very strong cider.

The tree is very hardy and bears profusely, so the crop is usually very heavy though the fruit is so small. The apples often hang on the trees like ropes of onions. It is a sure bearer every other year, and the fruit keeps well.

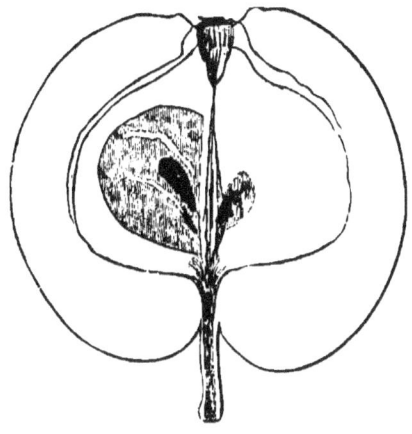

TANNER'S RED.

This apple is without any known history. It is very much grown, and seems to have originated in the neighbourhood of Canon Pyon.

Description.—Fruit: below medium size, oblong. Skin: more or less red on the whole surface, but much more deeply so on the side next the sun, and everywhere marked with thin stripes of a deeper colour. Eye: closed, set in a small and very puckered basin. Stalk: slender, half an inch long, and set in a narrow cavity, which is usually lined with russet. Juice: plentiful, rosy amber, subacid, with some astringency.

The chemical analysis of the juice of the *Tanner's Red* (season 1883), by Mr. G. H. With, F.R.A.S., F.C.S., Trinity College, Dublin, gave the following results:—

Density of fresh juice	1·060
Ditto after 24 hours' exposure to air ...	1·060
100 parts of juice by weight, yielded of	
Sugar	11·424
Tannin, Mucilage, Salts, &c.	2·176
Water	86·400

The tree is hardy, and bears an abundance of late-keeping fruit.

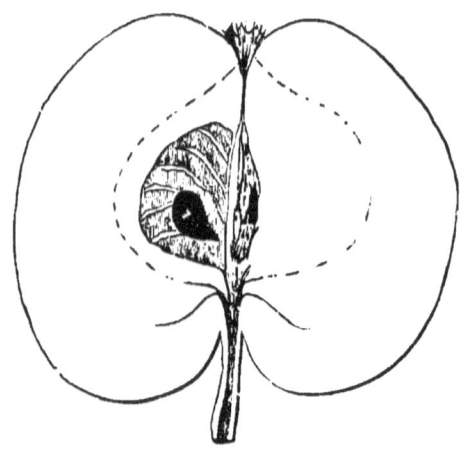

UPRIGHT REDSTREAK.

The upright habit of growth of the tree gives this variety its distinctive name.

Description.—Fruit: medium size, roundish oblate, regular in shape, but often fuller on one side. Skin: yellowish green, having a pink blush next the sun, with streaks and splashes of a deeper colour. Eye: small and closed, set in a small and shallow cavity. Stalk: slender, half-an-inch long, inserted in a deep and narrow cavity. Flesh: tender and juicy, with a sweet acidulated taste, and some astringency. Juice: plentiful, of a pale straw colour.

The chemical analysis of the juice of the *Upright Redstreak* (season 1881), by Mr. G. H. With, F.R.A.S., F.C.S., Trinity College, Dublin, gave the following results:—

Density of fresh juice ...	1·050
Ditto after 24 hours' exposure to air	1·050
100 parts of juice by weight, yielded of	
Sugar	12·280
Tannin, Mucilage, Salts, &c. ...	2·320
Water	85·400

This analysis proves it to be a valuable apple, which will make cider of excellent quality without mixture with other varieties, though it is seldom used alone.

There are many trees from 80 to 100 years old scattered throughout the orchards in the valley of the river Froome, and it is still propagated.

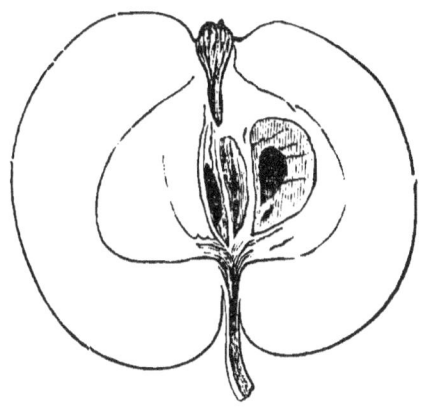

WHITE HEREFORD.

[SYN: *White Norman.*]

The origin of this apple is not known, but it is widely spread throughout the orchards.

Description.—Fruit: small, roundish, with obtuse angles on the sides, which are sometimes rather prominent. Skin: white, or rather a very pale straw colour, clear and waxlike, and with only a few large russet dots, distinctly sprinkled over the surface; the stalk cavity is lined with russet, which extends in ramifications over the base. Eye: very small, with narrow convergent segments, set in a deep basin, which is plaited, or slightly ribbed; tube, deep and conical; stamens, marginal. Stalk: long and very slender, deeply inserted. Flesh: snow white, soft and spongy, with a marked astringency, and bitterness mixed with sweetness. Cells of the core, open and very large for the size of the fruit. Cell wells, elliptical.

The chemical analysis of the juice of the *White Hereford*

(season 1878), by Mr. G. H. With, F.R.A.S., F.C.S., Trinity College, Dublin, gave the following results:—

Density of fresh juice	1·040
Ditto after 24 hours' exposure to air	1·042
100 parts of juice by weight, yielded of	
Sugar	10·770
Tannin, Mucilage, Salts, &c.	3·633
Water	85·597

This early variety yields a dark coloured cider, with a rich but slightly bitter flavour.

The tree is of middle size, and vigorous. It blossoms the beginning of May, and ripens its fruit by the end of October. It is very fertile.

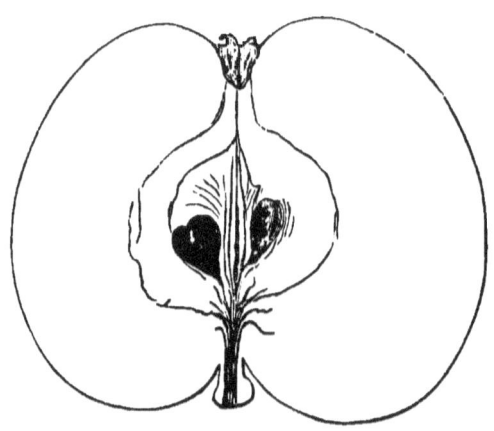

WHITE MUST.

[SYN : *White Musk.*]

This apple is a very old variety. It is mentioned by Evelyn

as "a great bearer, and its cider early ripe,"—and Phillips says of it :—

> "But how with equal Numbers shall we match
> The *Musk's* surpassing Worth! that earliest gives
> Sure hops of racy Wine, and in its Youth,
> Its tender Nonage, loads the spreading Boughs
> With large and juicy Offspring, that defies
> The Vernal Nippings, and cold Syderal Blasts!"

Description.—Fruit : roundish or oblate, even and regular in its outline. Skin: smooth and shining, of an uniform pale straw colour, which is a little deeper where it is more exposed to the light. Eye : small and open, set in a narrow and rather deep basin, which is round and smooth; segments, divergent; tube, short conical; stamens, basal. Stalk : short, and almost entirely within the cavity, and from which issues a ramifying patch of rough scaly brown russet, extending over the base. Flesh : yellowish, very tender, juicy, and pleasantly subacid. Cells of the core, closed; cell-walls, obovate. This is a pretty apple, and, after being gathered, its skin becomes quite unctuous, and gives off a powerful ethereal odour.

The chemical analysis of the juice of the *White Must* (season 1878), by Mr. G. H. With, F.R.A.S., F.C.S., Trinity College, Dublin, gave the following results :—

Density of fresh juice	1·037
Ditto after 24 hours' exposure to air ...	1·040
100 parts of juice by weight, yielded of	
Sugar	8·030
Tannin, Mucilage, Salts, &c. ...	3·580
Water	88·390

The *White Must* apple still retains its useful qualities, and is largely grown in all the cider counties of England. It produces a deep-coloured, sweet, and pleasant cider; but it has no great strength, and will not keep long.

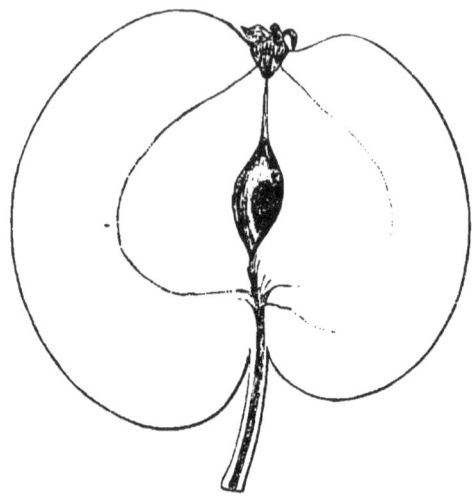

WHITE STYRE.

An old variety, widely scattered throughout Herefordshire and Worcestershire, but it is without any known history.

Description.—Fruit: about middle size, round, but obtusely ribbed. Skin: of a uniform lemon colour, with patches and lines of russet over the surface, especially on the side next the sun, and in the stalk cavity; the surface generally is strewed with small russet dots. Eye: closed, with connivent segments, set in a rather deep depression; tube, short funnel shaped; stamens, median. Stalk: slender, half an inch long, set in a deep russety cavity. Flesh: yellowish, soft and tender. Juice: plentiful, moderately sweet, and with a delicate sub-acid flavour. Cells of the core, open.

The chemical analysis of the juice of the *White Styre* (season 1880), by Mr. G. H. With, F.R.A.S., F.C.S., Trinity College, Dublin, gave the following results:—

Density of fresh juice ...	1·033
Ditto after 24 hours' exposure to air	1·036
100 parts of juice by weight, yielded of	
Sugar	9·100
Tannin, Mucilage, Salts, &c.	3·500
Water	87·400

This apple was formerly highly esteemed amongst the early varieties of cider fruit in Herefordshire, and is still valued in Worcestershire. It makes a light pleasant cider, of a deep colour, with good keeping qualities, but it is without much flavour, and with but little alcoholic strength. The fruit is therefore seldom used alone.

The tree is very hardy, and flowers the middle of May. It bears abundantly, and seldom fails to bear. The sandy loams of Worcestershire, with the blue clay (Lias) sub-soil, seems to suit it better than the strong clay loams of Herefordshire. The variety is old, and but very little propagated now.

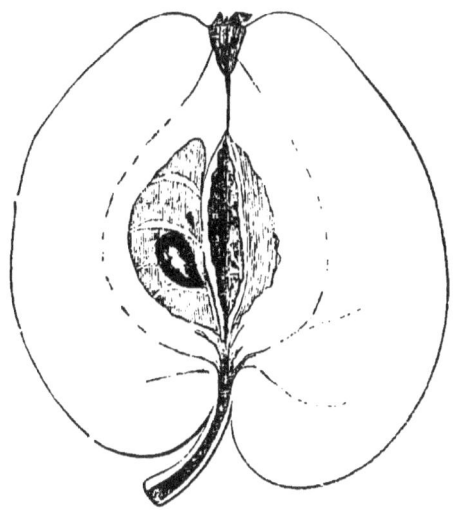

WILDING BITTER-SWEET.

A *Wilding* that has made its way by its own merit; a variety without any definite history.

Description.—Fruit: roundish ovate, often conical and ribbed, exactly of the shape, and very similar to the *Keswick Codlin*. Skin: pale yellow, tinged with green, strewed with russet dots, which have sometimes a greenish tinge surrounding them. Eye: small and closed, with converging segments, and set in a narrow, ribbed basin. Stalk: short, inserted obliquely by the side of a prominent lip, in a narrow, shallow cavity. Flesh: white and tender. Juice: moderate in quantity, of a deep amber colour, and of a vapid bitter-sweet flavour.

The chemical analysis of the juice of the *Wilding Bitter-sweet* (season 1881), by Mr. G. H. With, F.R.A.S., F.C.S., Trinity College, Dublin, gave the following results:—

Density of fresh juice ...	1·038
Ditto after 24 hours' exposure to air	1·040
100 parts of juice by weight, yielded of	
Sugar	10·420
Tannin, Mucilage, Salts, &c.	1·580
Water	88·000

The *Wilding Bitter-sweet* makes a high coloured sweet cider, and is now being propagated to some extent in the valley of the Frome.

The tree is hardy, grows freely, and bears well.

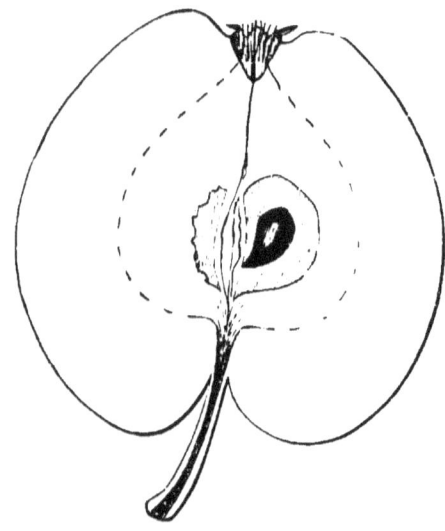

YELLOW REDSTREAK.

This variety is named from the colour of its fruit, a pale yellow ground-colour with thin crimson streaks.

Description.—Fruit: below medium size, often irregular in

shape. Skin: yellow, with faint streaks of red on the side next the sun. Eye: small and open, set in a narrow plaited cavity, segments slightly converging and then reflexed; anthers projecting. Stalk: slender, three-quarters of an inch long, inserted in a narrow and deep cavity. Flesh: tender, sweet and pleasant flavoured, with some astringency. Juice: very pale in colour.

The chemical analysis of the juice of the *Yellow Redstreak* (season 1881), by Mr. G. H. With, F.R.A.S., F.C.S., Trinity College, Dublin, gave the following results:—

Density of fresh juice ...	1·050
Ditto after 24 hours' exposure to air	1·053
100 parts of juice by weight, yielded of	
Sugar	12·380
Tannin, Mucilage, Salts, &c.	1·650
Water	88·970

The large amount of sugar contained in this fruit renders it valuable to mix with other varieties, which possess a greater amount of Tannin, Mucilage, and Salts.

The tree is hardy, grows to a good size, and bears freely.

AN ALPHABETICAL LIST

OF THE

MOST ESTEEMED VARIETIES OF
PERRY PEARS.

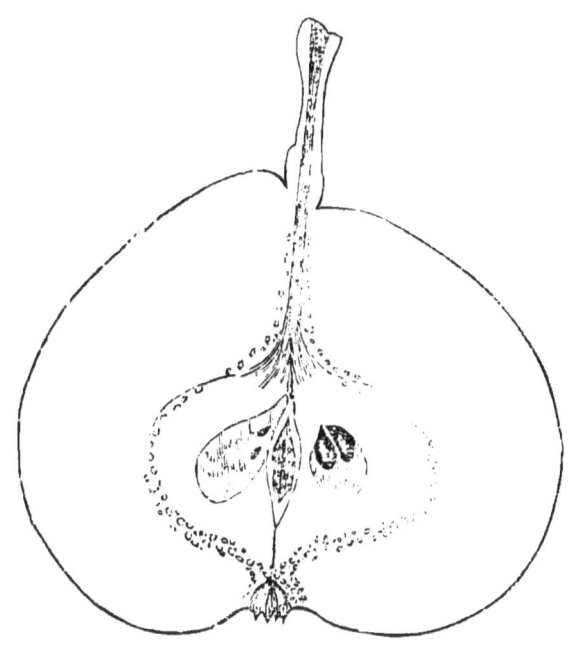

ARLINGHAM SQUASH.

[SYN: The *Green Squash* of Evelyn.]

This pear is undoubtedly a very old variety, and probably originated and took its name from the village of Arlingham, in Gloucestershire. This parish is formed by a nook of land surrounded on three sides by the river Severn, opposite Newnham. It has a rich alluvial soil, and many very old and large trees of the *Arlingham Squash* pear formerly grew there. Some few of them are still remaining, "all grafted," says Mr. Sayer, "by a single scion, and about the years 1700 to 1780;" this gentleman believes it to be the same as the *Green Squash* pear mentioned by Evelyn.

Description.—Fruit: a full medium size, roundish, almost *Bergamotte* shaped, but more irregular and lumpy. Skin: of a deep green colour, with a tinge of faint light red on the sunny side; the surface strewed all over with small brown spots of russet, and with patches of deep brown russet round the eye, the insertion of the stalk, and here and there about the body of the pear. Eye: an open ring, with traces of erect segments. Stalk: half an inch long, very thick, and enlarged at the base. Flesh: coarse and gritty, with an abundant juice of a deep amber colour, with a delicate sub-acid flavour and sweet taste, but followed by an astringent after-taste.

The chemical analysis of the juice of the *Arlingham Squash* (season 1881), by Mr. G. H. With, F.R.A.S., F C.S , Trinity College, Dublin, gave the following results:—

Density of fresh juice	1·039
Ditto after 24 hours' exposure to air	1·039
100 parts of juice by weight, yielded of	
Sugar	10·700
Tannin, Mucilage, Salts, &c.	1·800
Water	87·500

This rough-looking, ugly pear, is much esteemed by some growers. It is in season in early Autumn, and requires peculiar treatment. It is not fit to grind, until the inside is apparently rotten to within half an inch of the rind, when it "squashes" readily under the foot. The perry is sweet and good in flavour, but is only fit for immediate consumption. Wasps and bees are very fond of the decaying fruit, thus affording practical testimony to its aroma and sweetness.

The tree grows large and fine, and bears well. It is only grown in certain localities. A young and flourishing orchard of half-grown trees of this variety exists at this time at Bartestree, and other trees have been grafted from these, some fifteen or sixteen years since (1866-7) at Brockhampton.

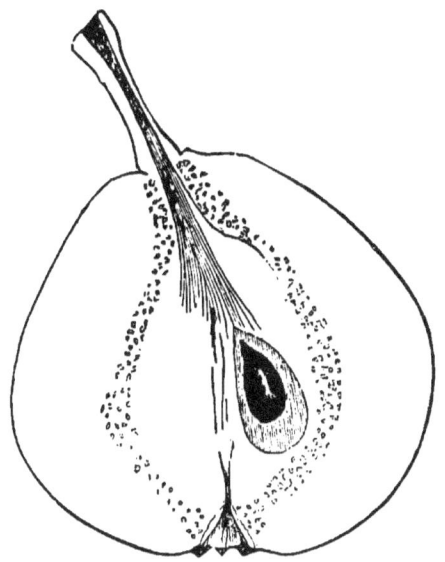

AYLTON RED.

[SYN: *Sack Pear; Black Horse Pear.*]

This pear seems to have originated in the hamlet of Aylton, about three miles west of Ledbury. It dates probably from the beginning of the present century, since none of the trees are old.

Description.—Fruit: middle size, roundish, turbinate, unequal in shape, being larger on one side than the other, with obtuse angles on the sides. Skin: pale green, red on the sunny side, with patches of deep crimson; its whole surface being strewn with russet dots, and with patches of thin russet. Eye: small, with thin reflexed segments, and sunk in a narrow puckered cavity. Stalk: short, half to three quarters of an inch long, stout, enlarged at both ends, and inserted in a narrow and shallow depression. Flesh: white. Juice: very plentiful, thin, of a pale amber colour, with a sweet taste, and an agreeable flavour, without much astringency.

The chemical analysis of the juice of the *Aylton Red* Pear

(season 1881), by Mr. G. H. With, F.R.A.S., F.C.S., Trinity College, Dublin, gave the following results:—

Density of fresh juice	1·036
Ditto after 24 hours' exposure to air	1·039
100 parts of juice by weight, yielded of	
Sugar	9·200
Tannin, Mucilage, Salts, &c.	4·000
Water	86·800

The *Aylton Red* Pear makes a rough perry, but without sufficient delicacy and richness of flavour, and sweetness, to gain for it a high character. It is better to be used mixed with other varieties.

The trees are not large, and are thin in foliage, but are very hardy. They blossom, bear freely, and ripen their fruit the middle of October. When laden with clusters of red, rosy fruit, as is commonly the case, the tree makes an attractive object in the orchard.

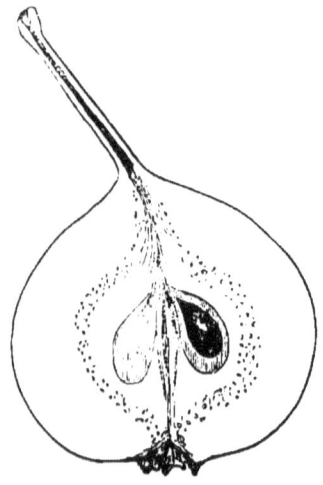

BARLAND PEAR.

[SYN: *Bosbury Pear; Bareland, or Bearland Pear.*]

" What tho' the pear-tree rival not the worth
Of *Ariconian* Products? yet her freight
Is not contemn'd.
    *   *   *   *   *   *   *
Chiefly the Bosbury, whose large increase
Annual in sumptuous banquets claims applause."

PHILLIPS' *Cyder.*

This pear, from one of its common names, may be supposed to have originated in the parish of Bosbury, near Ledbury, Herefordshire. The original tree is said to have grown in a field called *Bare Lands*, on an estate called "Bosbury Farm," and to have been blown down about the end of last century. The variety was well established in the 17th century, and in great repute. Evelyn (1674) says of it, "The *Horse Pear* and the *Bare-Land Pear* are reputed of the best, as bearing almost their weight of spriteful and vinous liquor. They will grow in common fields, gravelly and stony ground, to that largeness, as only one tree has been usually known to make three or four hogsheads." *(Evelyn's Pomona.)*

This fruit is well represented in Mr. Thomas Andrew Knight's "*Pomona Herefordensis*," Plate xxvii.

Description.—Fruit: small, turbinate, pinched in near the stalk. Skin: bright green, very much covered with patches and large dots of thick, pale brown, or ash grey russet, but not so much so, as entirely to obscure the green ground colour. Eye: large, for the size of the fruit, open, with short erect segments, filled with the permanent stamens. Stalk: an inch long, slender, and inserted in the end of the fruit, without any depression.

This variety has been much planted in Herefordshire and the adjoining counties. The trees have acquired an extraordinary size and height, and they are much distinguished by the beauty of their form and foliage. The largest orchards of this variety are now to be found in the parishes of Dymock, in Gloucestershire, and Newland, in Worcestershire. Very few farms on the eastern side of Herefordshire are without *Barland* pear-trees, showing how extensively this favourite variety was at one time cultivated. Evelyn several times mentions the *Barland Pear*, "and as no trees of this variety," says Mr. Knight, "are found in decay from age, in favourable soils, it must be concluded that the identical trees which were growing when Evelyn wrote, still remain in health and vigour. The specific gravity of the juice is 1·070."

The fruit, Evelyn describes as "of such insufferable taste, that hungry swine will not smell to it, or if hunger tempts them to taste, at first crash, they shake it out of their mouths:" but of the perry he speaks much more favourably. "There is a Pear in Bosberry and that neighbourhood, which yields the liquor richer the second year than the first, and so, by my experience, very much amended the third year." Another writer says: "It hath many of the Masculine Qualities of Cyder. It is quick, strong, and heady, high coloured, and retaineth a good vigour . . . many years before it declineth . . . As it approacheth to the Apple Cyder in Colour, Strength and excellence in Durance, so the bloom cometh forth of a damask Rose Colour, like Apples, not like other Pears."—*Herefordshire Orchards*, by J. Beale (1760).

The juice is rich in colour and full in flavour, its chemical analysis by Mr. G. H. With, F.R.A.S., F.C.S., Trinity College, Dublin, gave the following results :—

Density of fresh juice ...	1·0421
Ditto after 24 hours' exposure to air	1·0435
100 parts of juice by weight, yielded of	
Sugar	10·670
Tannin, Mucilage, Salts, &c. ...	2·763
Water ...	86·567

Mr. Knight, in his Pomona, says: "many thousand hogsheads of Perry are made from this fruit, in a productive season; but the Perry is not so much approved by the present, as it was by the original planters. It, however, sells well, whilst new, to the merchants, who have probably some means of employing it with which the public are not acquainted; for I have never met with it more than once, within the last twenty years, out of the district in which it is made, and many Herefordshire planters have applied to me in vain, for information respecting its disappearance. It may be mingled in considerable quantity with strong new port, without its taste becoming perceptible; and, as it is comparatively cheap, it possibly, sometimes, contributes one of the numerous ingredients of that popular compound."

Barland Perry does not bottle well. It curdles in the bottles. It is usually drunk, in Herefordshire, as soon as made, when it is considered very wholesome, and singulary beneficial in nephritic complaints.

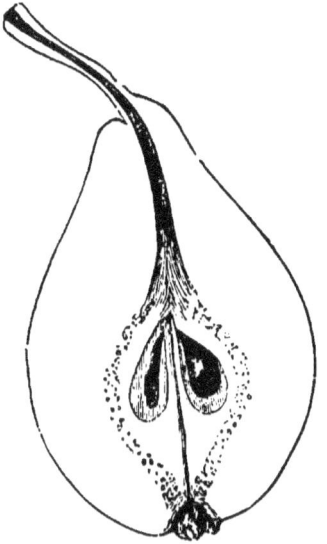

BLACK HUFFCAP.

[SYN: *Black Pear; Brown Huffcap.*

The *Huffcap* pears have been known from the 17th century. They were thought by Mr. T. A. Knight to have been included amongst the "Choke Pears" of the early writers, and which abounded in Herefordshire at that time.

This pear is represented in the 'Pomona Herefordensis," pl. xxiv., under the name of *The Huffcap Pear*, as the best of all the varieties.

Description.—Fruit: oblong, obovate, sometimes elliptical, tapering gradually from the bulge both towards the eye and the stalk; it is even and regular in its outline. Skin: olive green on the shaded side, and entirely covered with dull, rusty red, on the side next the sun; the whole surface being thickly sprinkled with large, grey russet dots. Eye: prominently set; open, with erect segments. Stalk: three-quarters of an inch long, woody, connected with the fruit by a thickened continuation of the flesh. Flesh: yellowish green, firm and very gritty.

The chemical analysis of the juice of the *Black Huffcap* (season 1879), by Mr. G. H. With, F.R.A.S., F.C.S., Trinity

College, Dublin, gave the following results:—

Density of fresh juice ...	1·048
Ditto after 24 hours' exposure to air	1·051
100 parts of juice by weight, yielded of	
Sugar	11·225
Tannin, Mucilage, Salts, &c.	3·575
Water	85·200

This fruit says Mr. Knight "is excessively harsh and austere, but it becomes very sweet during the process of grinding. Its Perry possesses much strength and richness, and has the credit of intoxicating more rapidly than that made from any other pear." It retains these characters at the present time and is highly esteemed.

The tree is hardy. It blossoms the beginning of May and bears abundantly.

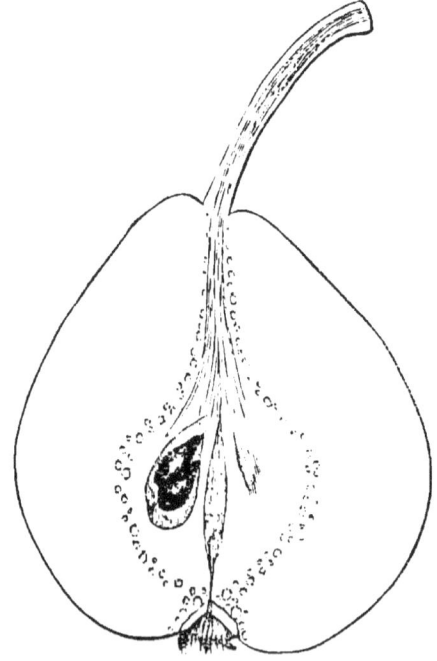

BLAKENEY RED.

There is no published history of this fruit. It may possibly derive its name and origin from the parish of Blakeney in the Forest of Dean.

Description.—Fruit: above medium size, of a regular pyriform shape. Skin: smooth, yellowish green on the shaded side, and a bright crimson on the side towards the sun, covered more or less with a very thin russet, and numerous very small spots. Eye: small and open, with upright calyx segments, set in a depression. Stalk: slender, an inch and a quarter long, set in a narrow cavity. Flesh: soft, moderately juicy, and very sweet, with a slight Jargonelle flavour, with very little astringency, but with a slightly bitter after-taste.

The chemical analysis of the juice of the *Blakeney Red Pear*, by Mr. G. H. With, F.R.A.S., F.C.S., Trinity College, Dublin, gave the following results:—

Density of fresh juice ...	1·033
Ditto after 24 hours' exposure to air	1·034
100 parts of juice by weight, yielded of	
Sugar	9·680
Tannin, Mucilage, Salts, &c. ...	3·160
Water	87·160

This coarse, showy pear, has, perhaps for these reasons, become lately very popular. It is saleable as a pot fruit for the manufacturing districts, but it is really a very worthless variety in the Orchard, and the sooner the large number of young trees planted within the last ten years, are grafted with varieties of higher merit, the better. The Perry made from its juice is rough and coarse in flavour, "abominable trash," and fit only for the most ordinary purposes, when nothing better can be got.

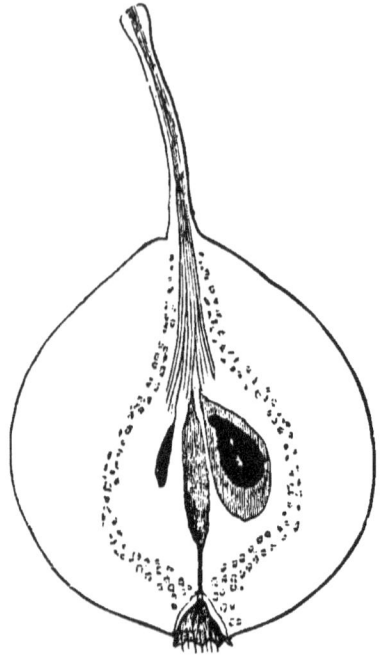

BUTT PEAR.

The origin of this pear is nowhere given. It is very much grown in Gloucestershire, on the Cheltenham side, and is spreading fast into Herefordshire and Worcestershire. Many of the trees are getting old, and the variety must therefore date from the last century, though it is not mentioned in the works of any of the orchard authorities.

Description.—Fruit: small and pyriform, elongated towards the stalk. Skin: of an uniform pale green colour. Eye: on the surface, with small erect segments, without much substance. Stalk: very slender, an inch long, inserted even, without depression on the narrow end of the fruit. Flesh: white and juicy. Juice: of full amber colour, not particularly sweet, and with a slightly bitter taste, and so much astringency as to roughen the palate very decidedly.

The chemical analysis of the juice of the *Butt Pear* (season

1881), by Mr. G. H. With, F.R.A.S., F.C.S., Trinity College, Dublin, gave the following results:—

Density of fresh juice	...	1·042
Ditto after 24 hours' exposure to air	...	1·044
100 parts of juice by weight, yielded of		
Sugar		10.700
Tannin, Mucilage, Salts, &c.		3·300
Water		86·000

This pear is becoming a great favourite in the orchards. It is in season very late, and is therefore the more useful, and the more easily managed. It makes a rough, strong Perry, which is at the same time sweet and good. It is often used to mix with other varieties to give strength to their Perry, whilst its own gains in softness.

The tree is very hardy and grows freely. It blossoms late, and rarely fails to bear abundantly.

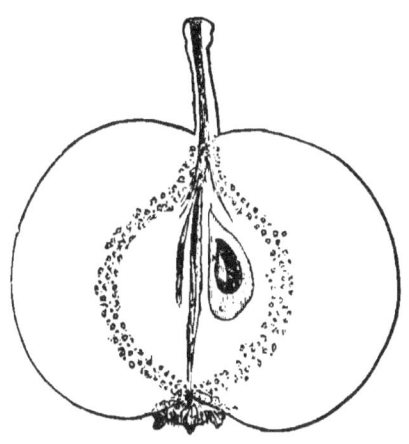

CHASELEY GREEN.

[SYN: *Hartpury Green.*]

This pear is believed to have originated in the parish of Chaseley, a scattered village in the district formerly called Malvern

Chase. It is also called *Hartpury Green*, from the village of Hartpury, in Gloucestershire, where it is much grown. It is without any known history.

Description.—Fruit : below middle size, two inches across and one inch and three-eighths high, round and flattened above and below. Skin: thick, of a fresh, pale green colour, becoming yellowish ; thickly studded with very distinct, thick, white, russet spots like scales. Eye : very open and shallow, with small upright segments set in a wide and shallow basin. Stalk: stout, from half to three-quarters of an inch in length, inserted without depression, but having often an irregular elevation of the fruit near it. Flesh : white, firm, more or less gritty. Juice : pale, mucilaginous, with a sweet, acid, and astringent flavour.

The chemical analysis of the juice of the *Chaseley Green* (season 1880), by Mr. G. H. With, F.R.A.S., F.C.S., Trinity College, Dublin, gave the following results :—

Density of fresh juice	1·047
Ditto after 24 hours' exposure to air	1·050
100 parts of juice by weight, yielded of	
Sugar	8·400
Tannin, Mucilage, Salts, &c.	5·600
Water	86·000

The fruit of the *Chaseley Green* pear, though capable of making a strong and rough perry, does not possess sufficient flavour to be used alone, except perhaps for home use. It resembles the *Holmer Pear* very much in shape, appearance and character, but is larger in size.

The tree has an upright growth until its boughs are bent down with the weight of fruit, for it is a prolific bearer. This pear is much grown in the lower valley of the Severn, both in Worcestershire and Gloucestershire; but it has only as yet crept into Herefordshire, in the neighbourhood of Ledbury, that is at Eastnor, the Homend and Eggleton, where there are many trees, and where it bears the name of *Hartpury Green*.

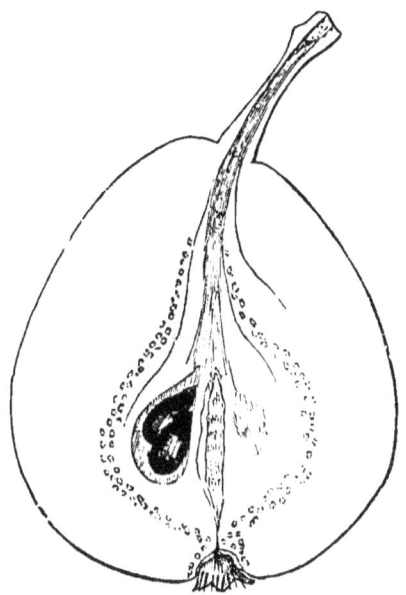

CHEAT BOY.

The history of this pear is not known. Its name indicates that its beauty is sometimes deceptive, and that the boys who purchase it are not to be congratulated on their bargain.

Description.—Fruit: pyriform, of medium size, tapering towards the stalk. Skin: greenish yellow, with a bright rosy colour on the sunny side, it has thin patches of russet round the stalk and eye, and spots of the same over the whole surface. Eye: small and open, level with the surface, calyx segments upright, anthers projecting. Stalk: slender, three-quarters of an inch long, inserted obliquely. Flesh: firm, juicy, sweet, with a slightly bitter after taste, with but little astringency. Juice: plentiful, of a pale amber colour.

The chemical analysis of the juice of the *Cheat Boy Pear* (season 1882), by Mr. G. H. With, F.R.A.S., F.C.S., Trinity

College, Dublin, gave the following results :—

Density of fresh juice	1·052
Ditto after 24 hours' exposure to air	1·052
100 parts of juice by weight, yielded of	
Sugar	12·700
Tannin, Mucilage, Salts, &c.	1·220
Water	86·080

An early variety, very pretty to look at, but of deceptive sweetness, with an unpleasant after-taste.

The trees are small, but bear very freely. It is a good pear, but not much propagated now. Trees of considerable age are to be found at Pendock, Berrow, Birtsmorton, &c., in Worcestershire.

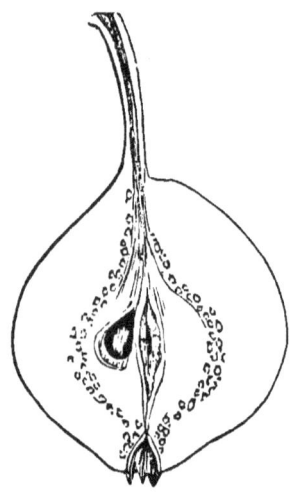

COPPY PEAR.

[SYN: *Coppice*.]

The origin of this pear is not known. There are many very large trees in Worcestershire, which proves the variety to be of great age.

Description.—Fruit: very small, growing in clusters, of a narrow ovate shape, with one side usually running up the stalk. Skin: yellow, and almost covered with small yellow russet spots. Eye: prominent, on a small ridge, with the projecting segments of the calyx closed. Stalk: slender, an inch or more long. Flesh: dry and pleasant to the taste, sweet and rich, with a great astringency. Juice: small in quantity, bright straw colour, becoming very dark on exposure to air, and somewhat viscid.

The chemical analysis of the juice of the *Coppy Pear* (season 1882), by Mr. G. H. With, F.R.A.S., F.C.S., Trinity College, Dublin, gave the following results :—

Density of fresh juice ...	1·057
Ditto after 24 hours' exposure to air	1·063
100 parts of juice by weight, yielded of	
Sugar	12·620
Tannin, Mucilage, Salts, &c.	4·380
Water	83·000

This analysis proves the pear to be more valuable than it is generally thought to be, since it is now no longer propagated. The more succulent varieties have taken its place in general estimation.

The trees are very large and spreading, with weeping slender boughs, very hardy, and so productive that the pears hang in bunches, and are therefore very small. There is a large orchard of this variety in the parish of Birtsmorton, Worcestershire.

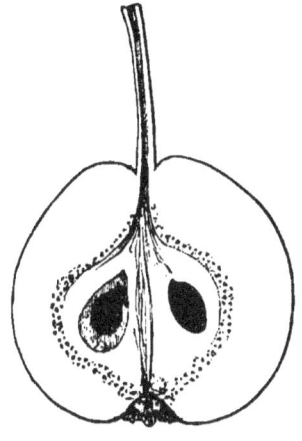

HOLMER PEAR.

[SYN: *Holmore*, by printer's error, in the *Pomona Herefordensis*.]

The original tree of this variety was found in a hedgerow, on the estate of Mr. Charles Cooke, of the Moor, in the parish of Holmer, near Hereford. Mr. Thomas Andrew Knight judged it to be about eighty years old (c. 1730), and the variety would now be about 150 years old. It is figured in the *Pomona Herefordensis*, Plate xx.

Description.—Fruit: small, roundish, turbinate, even and regular in outline. Skin: pale green at first, but of a dull greenish yellow, when ripe; thickly covered with russet dots, so as to form a kind of crust upon the surface. Eye: open, full of stamens, having short divergent segments, and set in a very shallow depression, or scarcely any depression. Stalk: from half to three-quarters of an inch long, slender, inserted in a small hole, with occasionally a slight swelling on one side. Flesh: yellowish, firm and crisp. Juice: plentiful, pale in colour, with a sweet, sub-acid, and very astringent flavour.

The chemical analysis of the juice of the *Holmer Pear*, (season 1882), by Mr. G. H. With, F.R.A.S., F.C.S., Trinity College, Dublin, gave the following results :—

Density of fresh juice	1·051
Ditto after 24 hours' exposure to air	1·055
100 parts of juice by weight, yielded of	
Sugar	11·900
Tannin, Mucilage, Salts, &c.	3·400
Water	84·700

Mr. Thomas Andrew Knight found the density of the juice to be 1·066, so that it possesses a high spirit-producing power.

The Perry from the *Holmer Pear*, in a good season, is of good flavour, sweet and rich, and resembles that made from the *Red Pear*.

The tree is strong and vigorous, grows tall, blossoms early in May, and bears well. The fruit follows the *Moorcroft* and *Barland* in season. The pears ripen altogether, and perish very quickly, so that they must be sent forthwith to the mill. It is not a favourite pear, but on Mr. Knight's recommendation of its rich juice, it was widely propagated, but this is no longer the case.

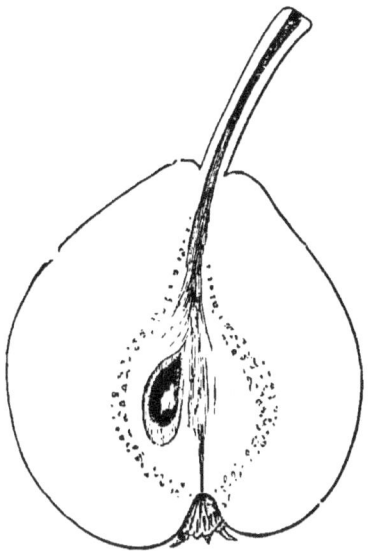

LONGLAND.

[Syn: *Longdon Pear.*]

The name of this pear, says Mr. Thomas Andrew Knight in the "*Pomona Herefordensis*," was probably derived from the field in which the original tree grew, but nothing is really known as to the circumstances or date of its origin. It is certainly a very old variety. This pear is well represented in the "*Pomona Herefordensis*," Plate xviii.

Description.—Fruit: roundish obovate, or doyenné-shaped, even, regular, and rather handsome. Skin: very thickly covered with large russet freckles of a pale ashy colour; the side next the sun has a bright, pale red cheek, and on the shaded side it is a greenish yellow. Eye: large, open, and clove-like, set even with the surface, with a ring of permanent stamens round the mouth. Stalk: an inch long, straight and stout, perpendicular with the axis of the fruit, and inserted in a narrow cavity. Flesh: yellow, very astringent.

The specific gravity of the juice Mr. Knight found to be 1·063.

The chemical analysis of the juice of the *Longland Pear* (season 1879), by Mr. G. H. With, F.R.A.S., F.C.S., Trinity College,

Dublin, gave the following results :—

Density of fresh juice	1·036
Ditto after 24 hours' exposure to air	1·041

100 parts of juice by weight, yielded of

Sugar	8·400
Tannin, Mucilage, Salts, &c.	4·187
Water	87·413

The trees of this variety are very hardy and productive, since the blossoms are extremely patient of cold and unfavourable weather. The Perry is very high coloured and without fine flavour. It is generally, however, free from sharp acidity, and more nearly resembles cider than any other kind of Perry. It does not answer for fining and bottling, but is excellent for ordinary use, either alone or mixed with apples, and its hardy, prolific character, makes it a general favourite.

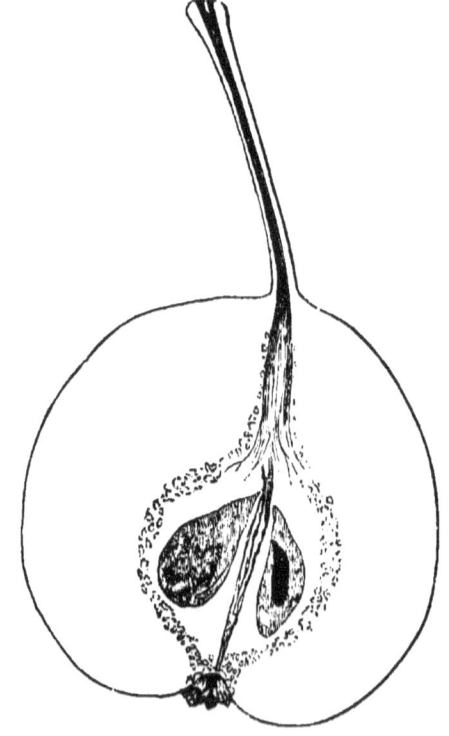

MOORCROFT.

[SYN: *Malvern Pear; Malvern Hill Pear.*]

This pear probably originated on the farm called "Moorcroft," in the parish of Colwall, near the western base of the Malvern Hills. There are many trees of considerable age there; it is chiefly cultivated in that district, and thus gets its synonym "*Malvern Hill Pear.*" Nothing however is positively known with regard to its origin.

Description.—Fruit: large for a Perry Pear, pyriform, even and regular in its outline. Skin: greenish yellow on the shaded side, becoming quite yellow as it ripens, with a brownish tinge on the side next the sun; the whole surface strewed with large ash-grey freckles of russet. Eye: open, set in a saucer-like basin. Stalk: half to three quarters of an inch long, rather stout, inserted without depression. Flesh: crisp. Juice: abundant, pale, with a sweet *Jargonelle* flavour, and some astringency.

The chemical analysis of the juice of the *Moorcroft Pear* (season 1880), by Mr. G. H. With, F.R.A.S., F.C.S., Trinity College, Dublin, gave the following results:—

Density of fresh juice ...	1·049
Ditto after 24 hours' exposure to air	1·050
100 parts of juice by weight, yielded of	
Sugar	11·916
Tannin, Mucilage, Salts, &c.	2·384
Water ...	85·700

This analysis proves that the *Moorcroft Pear* possesses a very rich juice, capable of making Perry of considerable alcoholic strength. It ripens very early, about the same time as the *Barland*, following the *Taynton Squash*, and before the *Red Pear* and *Oldfield*. The fruit is apt to decay soon, and care must be taken that it is used before this begins. It is usually mixed with other varieties, to impart to them its excellent flavour and sweetness.

The tree takes a spreading form of growth, attains a large size, blossoms the beginning of May, and is hardy, but can scarcely be called a free bearer.

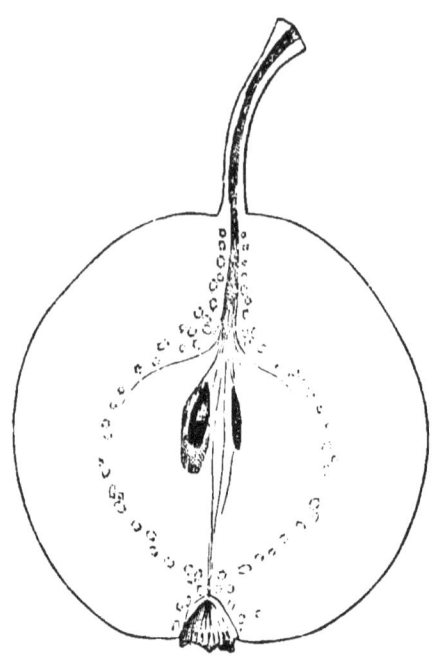

NEWBRIDGE PEAR.

A Worcestershire variety of considerable age, but, though the trees are large and handsome, and the variety popular, its history is unknown.

Description.—Fruit: full medium size, roundish oval, tapering a little towards the stalk. Skin: green, with a touch of orange brown on the side towards the sun, covered with minute russet spots, often large towards the eye, with a patch of thin russet round the stalk and eye. Eye: large, inserted on a level, with short upright calyx segments. Stalk: slender, nearly an inch long, inserted without depression. Flesh: very juicy, and sweet, with an astringent after-taste. Juice: plentiful, of a deep amber colour.

The chemical analysis of the juice of the *Newbridge Pear*

(season 1882), by Mr. G. H. With, F.R.A.S., F.C.S, Trinity College, Dublin, gave the following results :—

Density of fresh juice	1·049
Ditto after 24 hours' exposure to air	1·049
100 parts of juice by weight, yielded of	
Sugar	10·030
Tannin, Mucilage, Salts, &c.	2·670
Water	87·300

This is an early variety, and its Perry should be made in October, as soon as the first pear falls to the ground. The fruit is so very juicy that but little refuse is left in the hair bags. Twenty kipes of fruit will make a hogshead of clean drink. The Perry is very luscious and pleasant flavoured, of high colour, strong and clear.

The trees are very large and robust. The trunks are covered with bark as rugged and picturesque as the Spanish Chestnut, and the timber is well carried up into the branches. It is very hardy, and bears well. It is a very old and favourite sort. Some very fine trees are to be seen at Rye Court, Berrow, and in other Worcestershire orchards.

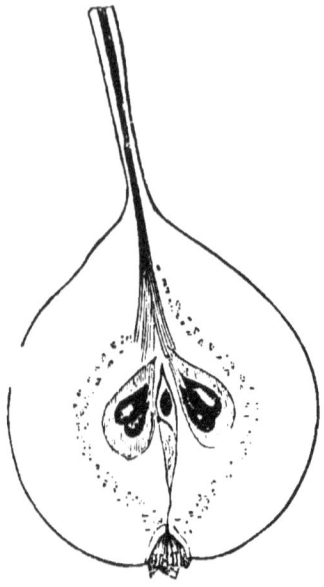

NEW MEADOW.

The origin of this pear, which is much grown around Ledbury, is not known, nor does its name give any clue to assist in the discovery.

Description.—Fruit: small and pyriform, almost entirely covered with a very thin russet, interspersed with many small, white spots. It has a light brownish green colour, with a tint of reddish orange next the sun. Eye: small, with upright segments, placed in a slightly depressed and puckered cavity. Stalk: slender, an inch long, red, with white spots on it, inserted on the surface of the fruit. Flesh: white. Juice: of a pale, pink colour, very sweet, with a pleasant aromatic flavour, recalling at first the dessert table, but followed by an after sensation of astringency.

The chemical analysis of the juice of the *New Meadow Pear*

(season 1881), by Mr. G. H. With, F.R.A.S., F.C.S., Trinity College, Dublin, gave the following results:—

Density of fresh juice ...	1·046
Ditto after 24 hours' exposure to air	1·048
100 parts of juice by weight, yielded of	
Sugar	12·000
Tannin, Mucilage, Salts, &c.	3·290
Water	84·710

The *New Meadow Pear* is not in great favour, because its juice is often very troublesome to carry through fermentation, and the Perry is also so peculiar in flavour, as not to be appreciated by everybody. As draught Perry from the cask, it is useful for home consumption.

The tree is hardy, upright and spiry in growth, and does not therefore waste much ground by its shadow. It is late both in blossoming and fruit bearing, and bears abundantly.

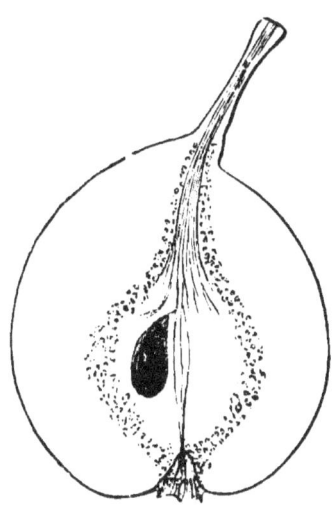

OLDFIELD.

This pear is believed to have derived its name from an enclosure called "Oldfield," near Ledbury, Herefordshire. There is no notice of it in any early catalogues of fruits. Phillips does not mention it, nor does it seem to have been known until the early part of the eighteenth century. An excellent figure of it is given in the *Pomona Herefordensis*, Pl. xi.

Description.—Fruit: small, round, even, and regularly formed. Skin: of a uniform greenish yellow when ripe, covered with minute dots, and a patch of russet round the stalk. Eye: open, with incurved segments, set in a shallow depression, surrounded with plaits. Stalk: an inch long, slender, not depressed, but tapering into the fruit at its base. Flesh: yellowish, firm, and crisp. Juice: pale, plentiful, sweet, and very astringent.

The chemical analysis of the juice of the *Oldfield Pear* (season 1880), by Mr. G. H. With, F.R.A.S., F.C.S., Trinity College, Dublin, gave the following results:—

Density of fresh juice ...	1·057
Ditto after 24 hours' exposure to air ...	1·061
100 parts of juice by weight, yielded of	
Sugar	13·060
Tannin, Mucilage, Salts, &c.	3·710
Water ...	83·230

Mr. Thomas Andrew Knight gives the density of this pear as 1·067, but states that it varies very much, like that of all other pears, according to the soil it grows on. The perry afforded by the *Oldfield Pear* is rich and sweet, with considerable strength, and ranks next to the *Taynton Squash* in general estimation. It fines readily in making, keeps well, and commands a high price in the market. It will keep and improve for 10 or 12 years in bottle.

The trees are large, very hardy, blossom the middle of May, and bear abundantly. The variety is very generally distributed throughout the county, and is in full luxuriance at this time.

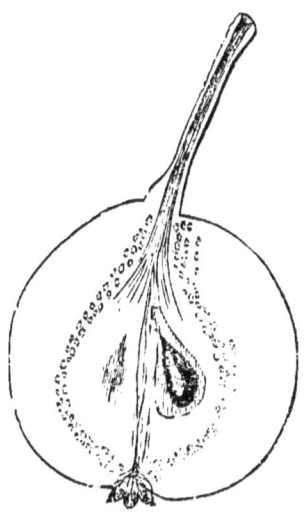

PARSONAGE.

The particular parsonage from which this pear takes its name, is lost to history, as is also any account of its origin.

Description.—Fruit : small and round, slightly running up the stalk. Skin: of a yellowish green colour, tinted with orange on the sunny side, and thickly strewn with very small russet dots. Eye : open, with erect segments, and placed on the surface, without depression. Stalk : slender, an inch long, enlarged at both ends, and set obliquely in the fruit. Flesh : white, with a sweet, astringent taste, and a slight *Jargonelle* flavour. Juice : plentiful, of a pale straw colour, deepening to amber, after exposure to air.

The chemical analysis of the juice of the *Parsonage* Pear (season 1881), by Mr. G. H. With, F.R.A.S., F.C.S., Trinity College, Dublin, gave the following results :—

Density of fresh juice	1·046
Ditto after 24 hours' exposure to air	1·052
100 parts of juice by weight, yielded of	
Sugar	9·600
Tannin, Mucilage, Salts, &c.	4·890
Water	85·510

The *Parsonage* pear gives an analysis which should render it more popular than it is, since it possesses such high keeping qualities. It is a very early pear, and its juice is rather deficient in sugar; and the perry is troublesome to make, from its liability to become ropy.

The trees are exceedingly large and upright, resembling very much the *Barland* in growth. They blossom the end of April, and the fruit ripens all at once, about the middle of September.

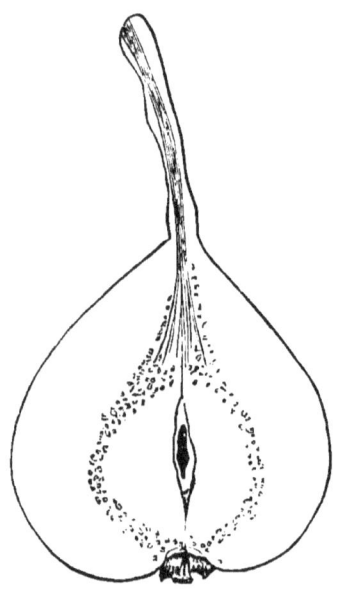

PINE PEAR.

The origin of this small pear is not known. It takes its name from its aromatic pine-apple flavour. The trees are old and large.

Description.—Fruit: small, of a flat, pyriform shape, broad below, and narrowing quickly towards the stalk, its sides being often unequal. Skin: green, with numerous very small spots on the surface, clustering together round the eye. Eye: open, and shallow, with upright segments, and set in a shallow basin. Stalk: very long, and very irregular in shape. Flesh: white, with a plentiful, thin, sweet juice, of a pale, amber colour, and vinous, pine-apple flavour.

The chemical analysis ot the juice of the *Pine Pear* (season 1881), by Mr. G. H. With, F.R.A.S., F.C.S., Trinity College, Dublin, gave the following results:—

Density of fresh juice	1·035
Ditto after 24 hours' exposure to air	1·040
100 parts of juice by weight, yielded of	
Sugar	9·300
Tannin, Mucilage, Salts, &c.	4·100
Water	86·600

The *Pine Pear* is generally confused with the next variety, the *Pint Pear*, but they are as distinct in character as in appearance. This variety is a late pear, and is generally used to give flavour to the juice from other pears. In a good year, when used alone, it makes Perry of a delicious flavour, and it bottles well. It is similar to the *Oldfield* Perry in flavour and character, and in its period of maturity.

The tree is hardy, grows to a large size, and bears well; but it is not generally grown, though it is still being propagated by those who know its value.

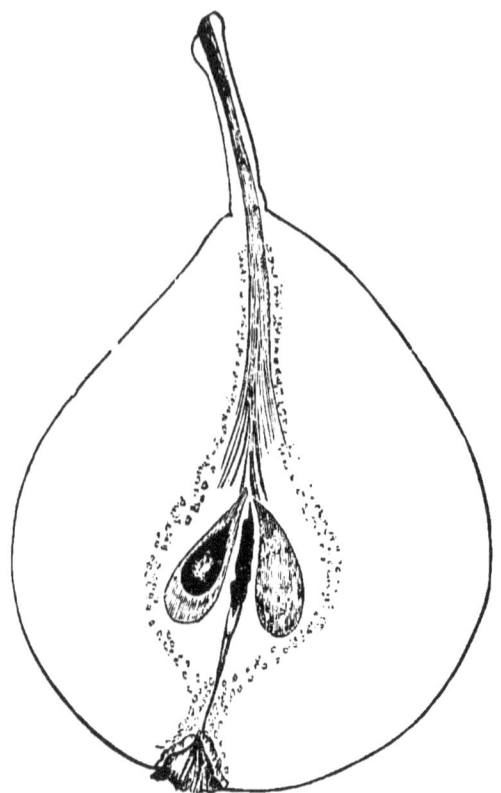

PINT PEAR.

Nothing is known of the origin of this pear. It may be inferred that its very juicy nature suggested its name. It is very much grown around Ledbury and in Worcestershire.

Description.—Fruit: of full middle size, with a regular, and rather round pyriform shape, tapering gradually towards the eye. Skin: of a pale green colour, with a slight tinge of orange on the side next the sun, and its surface is everywhere covered with very minute dots thickly placed. A thin cinnamon russet surrounds the eye and the insertion of the stalk, and is often seen in patches on the body of the fruit. Eye: small, and open, with erect segments, almost level with the surface. Stalk: three quarters of an inch long, inserted on the tapering end of the fruit without depression, but often with a fold of the fruit on one side of it. Flesh: white, and very juicy. Juice: very pale in colour, subacid, with a sweet, sharp, and rather astringent taste, though without any distinctive flavour.

The chemical analysis of the juice of the *Pint Pear* (season 1880), by Mr. G. H. With, F.R.A.S., F.C.S., Trinity College, Dublin, gave the following results :—

Density of fresh juice	1·039
Ditto after 24 hours' exposure to air	1·042
100 parts of juice by weight, yielded of	
Sugar	11·330
Tannin, Mucilage, Salts, &c.	1·370
Water	87·300

The *Pint Pear* from the abundance of its juice and the flavour of its Perry, has gained considerable favour. "It runs a lot of liquor," "it fills the measure well," are the usual observations first made about it. The Perry is apt to fret, and be troublesome in making, partly due to its being early ripe. Its Perry is pale in colour, and rough. It does not keep well, and should be drunk from the cask from Christmas to March. In this way, it is a very useful drink for home consumption, but its quality is not good enough for bottling, and thus it cannot take rank in the first order.

The tree is upright, very hardy, and free in growth. It blossoms the middle of May, seldom fails to bear well, and ripens its fruit the middle of September.

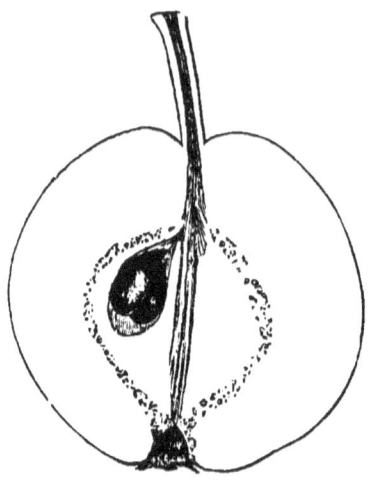

RED PEAR.

[SYN : *Red Horse Pear.*]

This Pear seems to have been well known in the 17th century, but its origin is involved in obscurity.

Description.—Fruit : small and round, even and regular in outline, but sometimes inclining to be turbinate. Skin : almost entirely covered with a rather bright, red colour, except round the stalk, and where it has been shaded, and there it is yellow; the whole surface is sprinkled with pale grey russet dots. Eye : open, having clove-like segments, and set level with the surface. Stalk : three quarters of an inch long, stout and straight with the axis of the fruit, set in a narrow, shallow cavity. Flesh : pale yellow, firm, dry, and gritty.

The chemical analysis of the juice of the *Red Pear* (season 1879), by Mr. G. H. With, F.R.A.S., F.C.S., Trinity College,

Dublin, gave the following results :—

Density of fresh juice ...	1·039
Ditto after 24 hours' exposure to air	1·039
100 parts of juice by weight, yielded of	
Sugar	8·742
Tannin, Mucilage, Salts, &c. ...	3·202
Water	38·056

The Perry from the *Red Pear* is very good, and has a strong cider like character. It bottles well, and in a good year makes a very saleable beverage.

The tree is very hardy. It blossoms the end of April, or the beginning of May, and generally ripens a heavy crop of fruit by the end of September, or the beginning of October.

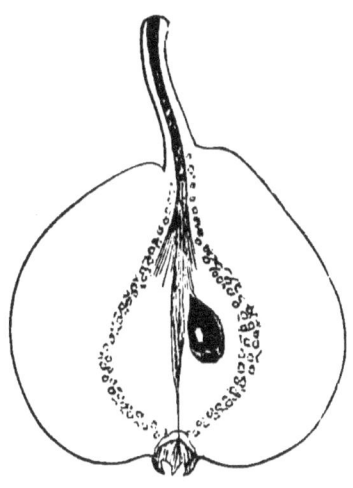

ROCK PEAR.

The original tree of this variety is still living at Cromer Pit Farm, Pendock, Worcestershire. It was raised in a little garden, enclosed from the waste, by an army pensioner named William

Tuffley, who attained the age of nearly 100 years, and the tree itself is now rapidly approaching that age.

Description.—Fruit: small and hard, irregular in size and shape, but usually roundish turbinate, flattened towards the eye, and larger on one side than the other. Skin: hard, of a dark green colour, with a reddish brown tint towards the sun, and spotted all over with minute spots of russet. Eye: small and closed, slightly depressed. Stalk: stout, half an inch long, and inserted in an irregular cavity. Flesh: hard, rough, acid, and astringent in taste, neither juicy, nor very sweet. Juice: full flavoured, of deep amber colour, viscid and very astringent.

The chemical analysis of the juice of the *Rock Pear* (season 1882), by Mr. G. H. With, F.R.A.S., F.C.S., Trinity College, Dublin, gave the following results:—

Density of fresh juice	1·075
Ditto after 24 hours' exposure to air	1·084
100 parts of juice by weight, yielded of	
Sugar	17·600
Tannin, Mucilage, Salts, &c.	4·150
Water	78·250

This analysis proves the very great value of this fruit. "It is a *Foxwhelp* amongst pears" said the analyst, without knowing anything of its character in the Orchard.

The Perry made from the *Rock Pear* has a rich, full, and rough flavour, of so much strength that it is said, "a man cannot drink enough of it to quench thirst, without incurring the risk of intoxication." Thirty-five or forty, three peck kipes of fruit, or in less local words, from twenty-six to thirty-two bushels, would be required to make a hogshead of clean Perry. In consequence of its great strength it is very rarely used alone, but it is a very valuable variety to mix with free running pears with an inferior quality of juice.

The trees are hardy, of good habit, and with well shaped heads of upright growth. The blossom is late, and a good crop of fruit may be looked for every other year with tolerable certainty. It is a very late pear, and the fruit will hang on the trees until all the leaves are down, if allowed to do so.

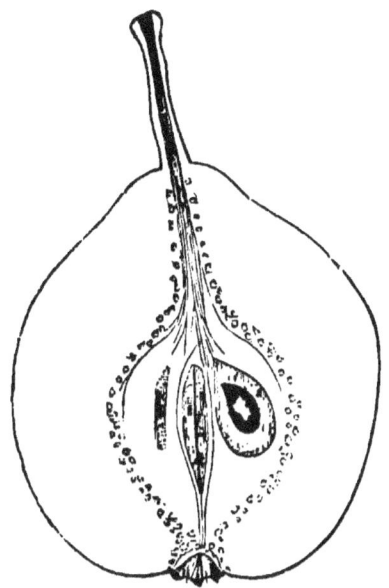

STONY WAY PEAR.

"Stony Ways" are not uncommon. This pear however is said to have originated at the Stony Way, near "The Winnings," at Colwall. It has only been brought into notice during the last twenty years.

Description.—Fruit: below medium size, of an irregular oval shape elongated towards the stalk, and usually bulging out on one side. Skin: greenish yellow, slightly tinted with brownish red towards the sun, and covered freely with minute spots of russet. Eye: small and open, in a small and shallow cavity. Stalk: slender, an inch long, inserted without depression. Flesh: firm and juicy, with a rather vapid taste, very slightly bitter, but without astringency. Juice: plentiful, of a pale straw colour.

The chemical analysis of the juice of the *Stony Way Pear* (season 1882), by Mr. G. H. With, F.R.A.S., F.C.S., Trinity

College, Dublin, gave the following results:—

Density of fresh juice	1·040
Ditto after 24 hours' exposure to air	1·042
100 parts of juice by weight, yielded of	
Sugar	8·360
Tannin, Mucilage, Salts, &c.	3·890
Water	87·750

The Perry is strong and of good quality, though wanting in sweetness. It is therefore seldom to be met with unmixed, and indeed, it may be said, that it is never made of this pear alone.

The trees are of round or spreading growth, but not large, and are said to be uncertain in bearing.

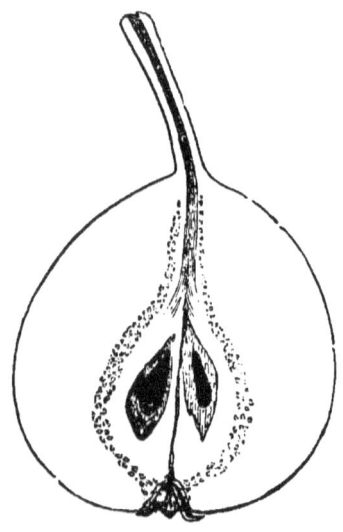

TAYNTON SQUASH.

"About Taynton (5 miles beyond Gloucester) Pears most abound, of which the best sort is that they name the *Squash Pear*, which makes the best Perry in those parts. These trees grow to be very large and exceedingly fruitful, bearing a fair round pear, red on one side and yellow on the other, when fully ripe, of a nature so harsh that Hogs will hardly eat them."

EVELYN, Pomona.

The earliest mention of this Pear is by Evelyn in the paragraph given above. There is no history of its origin, but its name and tradition lead to the belief that it is a native of the parish of Taynton. A *Red Squash* is mentioned by Worlidge, which may very possibly have been the same variety, since the great size and age of many of the trees sufficiently prove its antiquity. It is figured well in the *Pomona Herefordensis*, Pl. xiii.

Description.—Fruit : small, turbinate, even, and regular in outline. Skin : dull greenish yellow on the shaded side, and a clear red next the sun, with a few interrupted streaks of deeper colour ; a thin light brown russet runs more or less over the fruit, often in thickly clustered dots, but not sufficiently deep to mar its bright colour. Eye : open, with stiff, permanent, recurved segments, giving it a star like character, full of stamens, set in a shallow depression, and surrounded with plaits. Stalk: three quarters of an inch long, inserted without depression, with sometimes a fleshy lip on one side of it. Flesh : yellowish, abounding in juice of a rich, sweet flavour, brisk, and very astringent, but sometimes very disagreeably harsh and rough.

The chemical analysis of the juice of the *Taynton Squash Pear* (season 1880), by Mr. G. H. With, F.R.A.S., F.C.S., Trinity College, Dublin, gave the following results :—

Density of fresh juice	1·055
Ditto, after 24 hours' exposure to air ...	1·057
100 parts of juice by weight, yielded of	
Sugar	13·471
Tannin, Mucilage, Salts, &c.	3·033
Water	83·496

The *Taynton Squash* is the earliest of all the Perry Pears. the tree blossoms the end of April, and the fruit is ripe about the beginning or middle of September. It affords a Perry of the greatest excellence, with a sweet, rich, distinctive flavour, peculiarly its own. The *Taynton Squash* is among Perry Pears what the *Foxwhelp* is among Cider fruit, the first and the best. It is always sought after, and commands a high price.

The trees are hardy. They grow large and lofty with spreading branches. They bear freely. There is not a farm in Taynton parish without them, and they are scattered widely, but nowhere in great abundance. There are eleven trees on Aylestone Hill, Hereford; ten trees at Eggleton; and others scattered in less numbers, but many are dying from age. The Woolhope Club is successfully cultivating young trees.

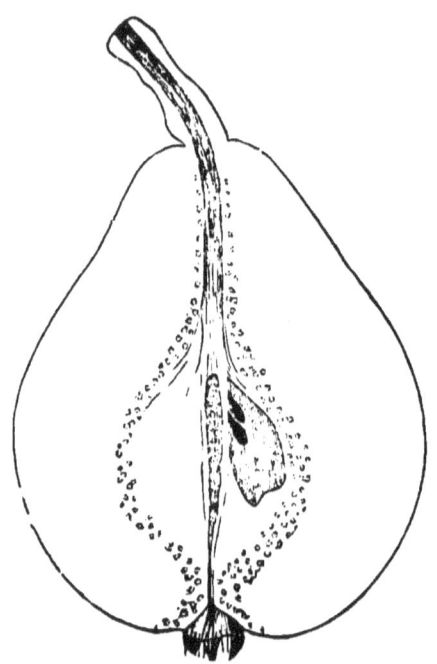

THORN PEAR.

An old variety without any known history.

Description.—Fruit: of full medium size or rather above it, of a blunt pyriform shape. Skin: of a light orange colour, with a crimson tint on the sunny side, and covered with thin russet specks over the surface. Eye: open, and scarcely depressed, calyx segments, incurved anther stiles, very long and erect. Stalk:

short, stout, and irregular, inserted rather obliquely, without depression. Flesh: firm, crisp and juicy, with an astringent after taste. Juice: plentiful, deep straw colour.

The chemical analysis of the juice of the *Thorn Pear* (season 1882), by Mr. G. H. With, F.R.A.S., F.C.S., Trinity College, Dublin, gave the following results:—

Density of fresh juice	1·046
Ditto after 24 hours exposure to air	1·048
100 parts of juice by weight, yielded of	
Sugar	11·500
Tannin, Mucilage, Salts, &c.	1·400
Water	87·100

The juice of the *Thorn Pear* makes a strong second-class perry —or cider, as the season may require. "A good useful liquor for home consumption," but it seems to require also a good country constitution to bear it, for "when fresh," the same informant added, "it will rout a body out well." It is a very early variety, and ripens all at once. As soon as a single pear falls to the ground, the fruit should be gathered and crushed. It is a very favourite pear in cottage gardens, for it stews well, and makes excellent pies and puddings.

The trees are small in size, and bushy, with stiff branches and large leaves. They bear too freely to make much wood. The trees bear so well, and the "fruit runs so much liquor," that its popularity in the gardens around Ledbury, and in Worcestershire, is very great.

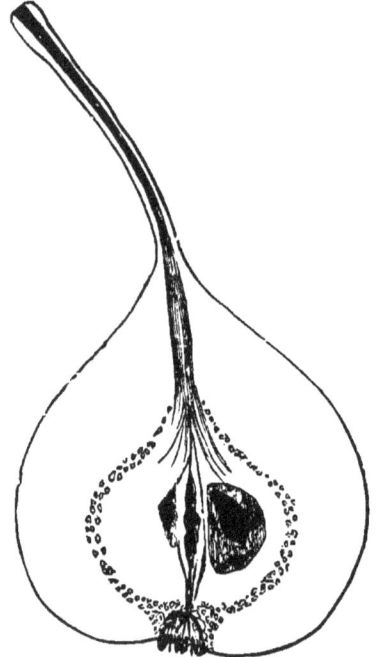

THURSTON RED.

[SYN : *Dymock Red.*]

The old family of the Thurstons held the estate of the Whitehouse, in the parish of Dymock for several generations. Mr. W. Thurston now lives there, and has several fine trees of *Thurston Red* Pear. He was told by his father that Mr. John Hiatt, formerly of the Merrables Farm, Dymock, a great fruit grower in his day, had grafted the young stock there, from the Whitehouse trees. This Whitehouse has the credit of being the birth-place of John Kyrle, the man of Ross. It is believed also to be the place in which this pear originated. Some eight or nine trees died there from old age; so the variety is ancient, though it has no history, and is now described for the first time.

Description.—Fruit : small, turbinate, even in its outline, but often fuller on one side than the other. Skin: smooth, greenish yellow, with a thin red cheek on the side next the sun ; it has often a large patch of thin, pale brown russet, especially round the eye, and a few spots here and there over the surface. Eye : small, and open, set in a saucer-like basin. Stalk : slender, an inch and a quarter long, set on the point of the fruit without depression. Flesh : yellowish and firm. Juice: thin, deep straw colour, sweetish, with an astringent, aromatic odour.

The Chemical analysis of the juice of the *Thurston Red* Pear (season 1880), by Mr. G. H. With, F.R.A.S., F.C.S., Trinity College, Dublin, gave the following results :—

Density of fresh juice ...	1·035
Ditto after 24 hours' exposure to air ...	1·036
100 parts of juice by weight, yielded of	
Sugar	9·200
Tannin, Mucilage, Salts, &c.	2·840
Water	87·960

This analysis is not favourable. It proves the juice to be thin and poor, and thus does not bear out the favourable character many growers give it. The fruit clings to the tree and keeps well, and hence is the more useful. It is a very local variety.

The tree is hardy with a nice upright growth, and bears well. It is cultivated extensively at Pauntly and Newent in Gloucestershire, and in the surrounding districts. Pauntly Court was long in the possession of the Whittingtons, from whom came the celebrated Dick, thrice Lord Mayor of London. Thus the *Thurston Red* Pear affects places of note.

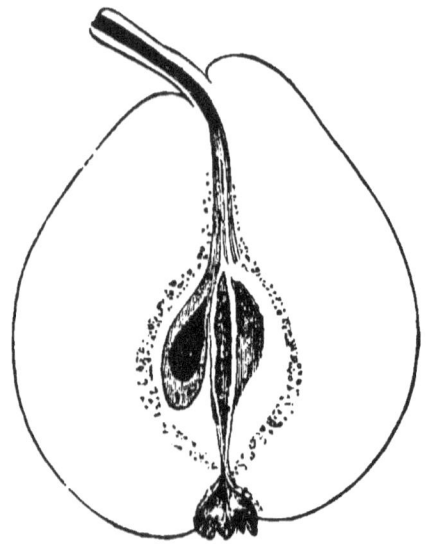

WHITE LONGLAND.

[SYN : *White Horse Pear.*]

The origin of this Pear seems to be unknown, but it is noticed by Dr. Beale in his "*Herefordshire Orchards*" (1657).

Description.—Fruit : oblong, obovate, even and regular in its outline. Skin : very thickly sprinkled with large russet dots and tracings of russet, and with a solid patch surrounding the stalk ; on the exposed side it has a thin pale red cheek, and on the shaded side, it is yellowish green. Eye : open, with short, incurved segments, set in a shallow depression. Stalk : half an inch long, woody, inserted in a narrow and shallow cavity. Flesh : yellowish, firm, coarse-grained, briskly acid and sweet.

The chemical analysis of the juice of the *White Longland* (season 1879), by Mr. G. H. With, F.R.A.S., F.C.S., Trinity

College, Dublin, gave the following results :—

Density of fresh juice	1·036
Ditto after 24 hours' exposure to air	1·039
100 parts of juice by weight, yielded of	
Sugar	8·580
Tannin, Mucilage, Salts, &c.	3·408
Water	88·012

"The *White Horse Pear*," says Dr. Beale, "yields a juice somewhat near the quality of Cyder." It is a favourite Pear in Herefordshire, not so much for its Perry—indeed it is seldom or never used alone for this purpose—as for its cooking qualities. It is an excellent baking pear, somewhat coarse, and rough in flavour, but with a natural deep rich red colour."

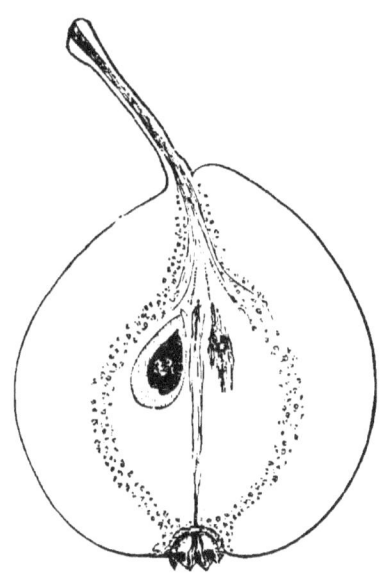

WHITE SQUASH.

[SYN : *Stanton*, or *Staunton Squash ; Squirt Pear.*]

The history and origin of this Pear is lost. Its synonym of *Stanton*, or *Staunton Squash*, may possibly indicate its origin to be a village of that name, between Ledbury and Gloucester, but there are other villages called Staunton.

Description.—Fruit : middle sized, turbinate, even, and regular in outline. Skin: yellowish green when ripe, and strewn all over with small russety dots, and here and there a patch of russet, and always russety round the stalk and the eye. Eye: open, with short, stunted segments, set in a saucer like basin. Stalk: an inch long, inserted without depression, and with a fleshy swelling on one side of it. Flesh: coarse and crisp. Juice: very abundant, of a deep amber colour, and harshly astringent.

The chemical analysis of the juice of the *White Squash* Pear (season 1880), by Mr. G. H. With, F.R.A.S., F.C.S., Trinity College, Dublin, gave the following results :—

Density of fresh juice	1·046
Ditto after 24 hours' exposure to air	1·048
100 parts of juice by weight, yielded of	
Sugar	10·611
Tannin, Mucilage, Salts, &c.	2·259
Water	87·130

This Pear is rich and sweet, but it quickly decays, and becomes, with a fair outside "rotten and squashy in the flesh." It makes a good family Perry if taken at the right moment, rich and sweet; but it is "stubborn to fine," and its readiness to run into watery decay, makes its power of cask filling, its chief merit.

The tree is of small size, but a great cropper. It is "lucky for bearing" they say, and thus it maintains its place in the orchard.

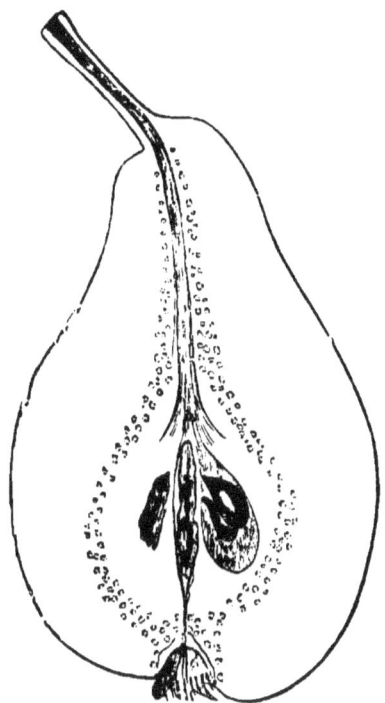

WINNALL'S LONGLAND.

This pear was raised by the late Mr. Winnnall, of Woodfield, in the parish of Weston-under-Penyard, near Ross, towards the close of the last century (c. 1790). The first orchard planted with it, was the one called "Noaks Style" on that estate, from whence his friends in Herefordshire, and Worcestershire, were supplied with grafts.

Description.—Fruit : handsome, rather above medium size, of long pyriform shape, tapering very much towards the stalk, but often fuller on one side than the other. Skin : greenish yellow, with a beautiful red tint, softened with bloom on the side exposed to the sun, the whole surface sprinkled with small spots, which become larger towards the stalk. Eye : small and open, set in a slight depression. Stalk: slender, nearly an inch long, often inserted obliquely. Flesh: soft, juicy and sweet, with a slightly bitter after taste, and without astringency. Juice: straw colour, very sweet, with a slight *Jargonelle* flavour.

The chemical analysis of the juice of the *Winnall's Longland* (season 1882), by Mr. G. H. With, F.R.A.S., F.C.S., Trinity College, Dublin, gave the following results:—

Density of fresh juice	1·045
Ditto after 24 hours' exposure to air	1·050
100 parts of juice by weight, yielded of	
Sugar	11·900
Tannin, Mucilage, Salts, &c.	1·780
Water	86·320

The perry is more luscious than that from the ordinary *Longland* pear, and very strong. It is rough in flavour, and not fit for bottling, but it is very saleable for ordinary purposes. With a little colouring from burnt sugar, it cannot readily be distinguished from cider, for which it is not unfrequently sold. It is difficult to make well, and in some districts is apt to get a smoky flavour.

The tree is very handsome in shape and park-like, of great size, and very vigorous. It bears profusely. Mr. Chas. Blandford, of Merrables, Dymock, said in 1880, "There are seven trees on my farm, with fruit enough on them to make 14 hogsheads of perry. During the five years I have been here, these seven trees have averaged 12 hogsheads of perry annually." It is also widely grown around Ledbury, and in Worcestershire, about Eldersfield, Birtsmorton, and other parishes in the valley of the lower Severn. It is still propagated extensively in these districts.

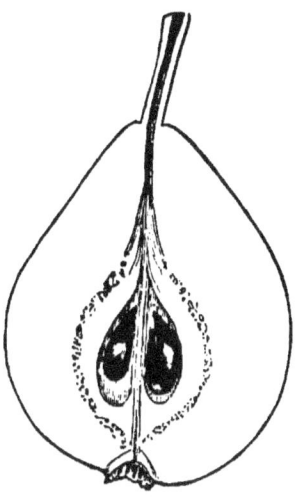

YELLOW HUFFCAP.

The history and origin of this Pear is lost. It is figured in the *Pomona Herefordensis* by Mr. Knight, Pl. xxiv.

Description.—Fruit: turbinate. Skin: entirely covered with rough, brown russet, but not so much so as to obscure altogether the green ground colour, which is shown through the specks. Eye: open, small, with short horny segments, set even with the surface. Stalk: three quarters of an inch long, inserted without depression. Flesh: yellowish, with a tinge of green.

The chemical analysis of the juice of the *Yellow Huffcap* (season 1879), by Mr. G. H. With, F.R.A.S., F.C.S., Trinity College, Dublin, gave the following results:—

Density of fresh juice ...	1·046
Ditto after 24 hours' exposure to air	1·049
100 parts of juice by weight, yielded of	
Sugar	11·244
Tannin, Mucilage, Salts, &c.	2·290
Water	86·466

The Perry of the *Yellow Huffcap* Pear is excellent. It is

richer, and has more body than the *Oldfield* Perry. " I always win the prize with the *Yellow Huffcap* Perry " says Mr. Hill, of Eggleton.

The *Yellow Huffcap* is a very favourite pear in the neighbourhood of Ledbury. The tree is very hardy. It blossoms the beginning of May, and bears every year, but usually in much greater abundance every second year.

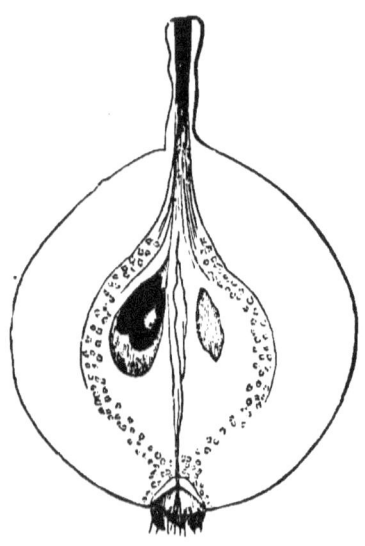

YOKEING HOUSE.

Another Worcestershire pear of unknown history.

Description.—Fruit : below medium size, turbinate. Skin : of a greenish yellow, scattered with russet, particularly around the eye and stalk. Eye : prominent, with long, projecting calyx segments, long anthers and pistils. Stalk : short and stout, about half an inch long, straight, or slightly oblique, with level insertion. Flesh : very sweet and juicy, with a pleasant aromatic taste, rich flavour, and very little astringency. Juice : of a pale straw colour.

The chemical analysis of the juice of the *Yokeing House* Pear (season 1882), by Mr. G. H. With, F.R.A.S., F.C.S., Trinity

College, Dublin, gave the following results :—

Density of fresh juice	1·060
Ditto after 24 hours' exposure to air	1·065
100 parts of juice by weight, yielded of	
Sugar	13·700
Tannin, Mucilage, Salts, &c.	2·300
Water	84·000

The Perry is pale or white in colour, sweet and good, but not strong. It is used to mix with other varieties.

The trees are of middle size, woody, and rather stiff in growth, with branches inclined to spread, rather resembling the growth of the Alder tree. Several trees are to be seen at Rye Court, Berrow, of a considerable size.

LOCAL PERRY PEARS.

There are many varieties of Perry Pears scattered through the orchards as single trees, or a very few together, which for the most part, are very coarse in their juices. They all bear well, and are allowed to remain because they are there, and are useful in filling the vats for home consumption. These varieties, however, have sometimes much local esteem, and it must be remembered, that it is from them, that experience points out the best varieties. The following names, and brief remarks from the note book, are the result of many visits to the Orchards :—

BOSBURY SCARLET.—A valuable pear of the mid-season. The tree bears abundantly, and its fruit makes excellent perry. It is being propagated very extensively in the Ledbury district.

TUMP PEAR.—An early variety, "too early to be of much use." It makes a strong rough sweet perry, of inferior flavour, which turns of a blackish colour on exposure to air.

FOREST PEAR.—Early, soft and juicy, used with other fruit. Tree, large and well grown, like an oak.

KNOCK DOWN.—A mid-season variety in the valley of the river Froome. It makes excellent perry, very like Barland. Is fined for bottling, and generally turns out well.

LONG STALK.—Makes excellent perry, as pleasant as sherry. Trees large, with fine limbs, as large as *Moorcroft*. Will grow 60 kipes (30 bushels) of fruit to a tree.

DYMOCK RED, AND TURNER'S BARN PEAR.—Two local varieties, in much repute near Ross.

GIN PEAR.—Very like *Barland*, supposed to be equally good for gravel, and hence perhaps its name.

LUMBERSKULL.—A great bearer. Makes a strong, rough perry, which turns dark coloured on exposure to air.

SOW PEAR.—A very late Worcestershire pear. Makes a rich strong perry, but not of agreeable flavour. A very old variety.

BLACK HORSE PEAR, AND WHITE HORSE PEAR.—Make a rough coarse perry, turning black on exposure to air.

VINTAGE FAVOURITE; WHITE MOORCROFT; GREGG PEAR; SACK PEAR; MILL PEAR; NORTON BUTT, &c., &c., &c., are other varieties, only known in their special localities.

LIST OF OTHER
CIDER APPLES,

FROM THE COUNTIES OF

HEREFORD, DEVON, SOMERSET, WORCESTER,
AND GLOUCESTER.

"Nec requies, quin aut pomis exuberet annus
Aut fœtu pecorum."
Virgil. Geor. II., 516—17.
The laded boughs their fruits in autumn bear.

"Quotque in floro novo pomis se fertilis arbos
Induerat, totidem autumno matura tenebat."
Geor. IV., 142—3.
For every bloom his trees in spring afford
An autumn apple was by tale restored.

The following varieties were exhibited at the Hereford Apple Shows, held under the auspices of the Woolhope Club in the years 1878, 1879, 1880, 1881, and 1883. They have not been described in the preceding pages. Their merits, for the most part, have not been accurately ascertained, though some of them are of excellent quality. They are placed alphabetically, with such observations about them, as have been obtained from the growers. Numerous other varieties have also been shewn, but they were alike wanting in name, history, and character.

ALFORD, or SWEET ALFORD.—A white Devonshire apple, of middle size. The tree is large and spreading, and bears freely. It is a late variety, has a sweet rich juice, and makes very good cider.

AMPHLETT'S FAVORITE.—A large striped apple, not unfrequently met with in Herefordshire Orchards, on the eastern side of the county. It is usually sold as table fruit in the market, but sometimes finds its way to the cider press as a cask filler.

ANSELL, or ANCELL.—A medium sized, red, russety apple, much grown at Oldbury, in the vale of Berkeley, Gloucestershire. The tree grows erect, and bears freely. It is a late keeping apple, and is highly esteemed for the excellent cider it makes.

BASTARD ROUGH COAT.—A long keeping russet apple, more fit for the dessert table than for cider making.

BAYLIS' KERNEL.—A streaked apple, of medium size and fair quality. It is ripe about midseason, and is thought to add good flavour to the cider from mixed fruit.

BELLE ORCHARD SEEDLING, or LEDBURY BELLE.—A middle sized apple, which in colour, shape and general appearance, somewhat resembles the *Foxwhelp*. The flesh is more or less red tinted, with good flavour and rich juice. It makes a cider of the first quality. The tree is of upright growth, very vigorous, and a very free bearer.

BENNETT APPLE.—An old variety figured by Mr. T. A. Knight. An orange striped apple of full medium size, and rather early. It has an abundance of sweet, rich juice, of the specific gravity 1·073 (Knight). It sells readily in the market as table fruit.

BEST BACHE, or BACHE'S KERNEL.—An old variety grown in Herefordshire orchards on the south eastern side of the county. It is of full medium size, with a broad base and angular sides. The colour is a rich yellow, streaked with pale and dark red; it has a rich juice of the specific gravity 1·073 (Knight), and is highly esteemed as a cider fruit.

BITTER-SCALE.—A Devonshire apple formerly held in high esteem. It does not however seem to have maintained its place in the orchard, and it is very doubtful if the fruit shewn was the true variety.

BLACK BUD, in contradistinction to *Red Bud.*—It is a dark red apple, of deep mahogany colour on the sunny side. It is chiefly grown in the valley of the river Froome. Its juice however is light and pale, and will help to fill the cider, or perry cask, with equal efficiency.

BLACK-EYED PIPPIN.- -A recent variety, much esteemed at Bishop's Froome, where it seems to have been raised. It makes a strong, full bodied cider, but not sweet enough for most people. It is most useful to mix with other varieties.

BLACK HEREFORD.—A large, white apple, grown in Somersetshire, and reported as "good for extra prime tipple." It is not known in Herefordshire, where the *Black Hereford* (formerly *Black Norman*) is a dark green apple, below middle size.

BLACK WILDING.—A fine looking conical apple, of a depressed colour, from the valley of the Froome. It is probably grown for its colour, since it has not much distinctive character as a cider apple.

BOTTLE STOPPER.—A Devonshire apple of good acidity, in high repute for apple jelly, and said also to make good cider.

BRIDGE PIPPIN.—An early Gloucestershire apple, yellow and sweet. The tree is erect and bears freely.

BRISTOL CRAB.—A cider or pot fruit. There are some fine trees at Moorcroft and Colwall. The fruit makes excellent cider, good enough to be "kept for the master's drinking."

BROAD-EYED PIPPIN.—A yellow apple somewhat like the *Downton Pippin*, but larger. It is second early, a fair dessert fruit, and should be sold as such, since it has no especial merit for the cask.

BROAD-LEAVED HEREFORD (formerly *Broad-leaved Norman*).— A large pale green apple, with a slight flush of red on the sunny side, of a conical shape with obtuse angles. The trees grow freely with large foliage, and they are all comparatively young trees, so the variety must be recent. They bear an abundance of fruit of a sweet and slightly bitter taste. It makes good cider.

BROADTAIL.—A very productive variety, which comes quickly into bearing. It is grown widely in the northern and eastern districts of Herefordshire. It is a hard fleshed apple, which keeps well and is often sold in the market as pot fruit. It is not a good apple however, and its cider is pale and without character.

BROMESBERROW CRAB.—An apple mentioned by Evelyn, and formerly in high repute. It is not now met with, and those shown as such did not answer to its old character.

BROWNSEYS.—A Somersetshire apple, large and striped. It is usually sold as a table fruit, but is often used as a cask filler.

BROWN SNOUT, OR POINTED BROWN SNOUT.—A very good late apple. The tree grows freely, and is a very heavy cropper. The apple is green in colour, and firm in substance. It has a projecting eye. It is a bittersweet, and makes good cider.

BULL'S EYE.—A red apple of medium size, hard in texture, and a late keeper. The tree bears well. There are many trees in the parish of Marden. They droop in growth, bear well, and the fruit is much esteemed for the quality of its cider.

CABBAGE APPLE.—A large green apple grown in Gloucestershire. The tree is erect, and bears well. It is a midseason apple, and is often sold as pot fruit.

CANON APPLE.—An apple of some repute at Canon Pyon. It is a pleasant looking fruit, but the examination of its juice was not satisfactory. It can only be classed as a cask filler, which requires body and flavour from other fruits.

CANDID HEART.—An apple above middle size, good either for cooking purposes, or for cider. It is a great and constant bearer.

CANON BITTER-SWEET.—A greenish, slightly streaked apple, of medium size. The tree bears freely. Its fruit is late in season, and is esteemed for its cider.

CAPTAIN NURSE, CAPTAIN'S KERNEL, or NURSE'S KERNEL.— A Gloucestershire apple, much streaked and coloured with red. The tree grows slowly, but bears well when full grown. It is a late variety, but has not much character as a cider fruit.

CHAXHILL RED.—A very beautiful little Gloucestershire apple, which received a first-class certificate at Gloucester (1873) "for its excellence as a cider fruit." It was raised from seed by Mr. Bennett, of Chaxhill, Westbury-on-Severn. Its juice, however, is poor and thin, and it has not therefore maintained its character as a cider apple.

CHIBBLE'S WILDING.—A yellow Somersetshire apple, highly esteemed as a cider fruit from the richness of its juice, and the briskness it is believed to impart to the cider. The tree bears well.

CIDER BRANDY APPLE.—A small, dark coloured apple, much grown in Worcestershire, where it is held in great repute. It is something like *Kingston Black*, but much softer in texture.

CLARET-WINE APPLE.—A deep purple tinted apple, whose chief merit is its colour.

COLEING.—"Grown about Ludlow," says Evelyn; but it is seldom heard of in these days.

COOK'S KERNEL.—A favourite apple in some districts of Herefordshire. It is above medium size, second early, or late. The tree grows large, and bears well. It is an excellent variety, and is said to make "the fullest mouth cider of any kind."

CORN APPLE, or HARVEST APPLE.—An early, red striped, conical apple, which makes a pleasant drink for hop-pickers. It has a sweet rough taste, and usually finds its way into the costermongers' carts. Its cider is only nice when drank as soon as made.

DARBIN RED STREAK.—A Somersetshire red streak, of much esteem for its cider.

DEAN'S APPLE.—A Devonshire apple of large size, which belongs rather to the table than to the cider press.

DEVONSHIRE RED STREAK.—An apple of middle size, good for cider or pot fruit. It is much grown in Worcestershire and Gloucestershire, and about Ledbury. The tree has a drooping habit, and bears well every, or every other year. Its fruit is mid season, and makes good cider.

DEVONSHIRE ROYAL WILDING, sometimes called the *Red Hill Crab*, from a hill on the highway on which the original tree grows. This variety is mentioned with the highest praise by Mr. Hugh Stafford, of Pynes (1753). He denotes it as a wilding, growing in "a little gillet of gardening" on the highway side, one mile from the city of Exeter, on the border of the parish of St. Thomas. "Sixteen years since" (i.e., 1737), he says "it was grafted very much by the Rev. Robert Woolcombe, Rector of Whitestone, the adjoining parish." Mr. Stafford was personally acquainted with Mr. Woolcombe, and learnt all the particulars from him. Mr. Woolcombe thought it so superior to all other apples for cider, that he gave it the name of *Royal Wilding*. The cider has great roughness and body. "I will venture to affirm," says Mr. Stafford, "that I have never tasted any cyder equal to it (not all the genuine Hereford I ever drank) that of the *Winesour* only excepted." He has known "five guineas refused for a hogshead of its cyder, whilst common cyder sells for twenty shillings, and South Hams from twenty to thirty." When cooked, he adds, "it has something of the rough flavour of the Quince." The *Devonshire Royal Wilding*, exhibited at the Hereford Apple Shows, was a larger table fruit, without the qualities denoted by Mr. Stafford; and the Committee tried in vain to procure the true variety from Devonshire.

DUFFLIN.—An old Devonshire apple, formerly much esteemed, but it is doubtful if the true variety is now to be found.

DUNN'S BELOVED.—A pretty, attractive apple. The tree bears freely. It is a good filler, but its juice is light in density; its cider is difficult to fine. It quickly turns a dark colour on exposure to air. The fruit keeps well, so it should be picked and sold as pot fruit in the early spring.

ESSEX KERNEL.—A very good, late cider apple. It is lemon-shaped and yellow, streaked with red. It is rough and russety around the eye and stalk. The tree is large and bears well, and the fruit makes excellent cider of a deep yellow colour.

EXCELS.—A pale, red streaked, second early apple. The tree is small in size, but crops well.

FARMER HEARLAND.—A Somersetshire apple of large size and yellow colour. The tree is upright in growth, and bears a fruit that keeps well, and is said to make good cider.

FAWKES' KERNEL.—An apple above middle size, with a broad base and irregular sides. The eye is deeply sunk. The skin is thick, of a pale yellow colour, becoming orange on the sunny side, and with numerous minute, dark, point-like spots scattered over the surface. The fruit yields a cider of high quality, and sells readily also for kitchen purposes. The trees grow freely to a large size, and crop well. It is a valuable variety, much grown about Dymock, Ledbury, and occasionally elsewhere.

FILLETS or VIOLETS, SUMMER and WINTER.—Apples formerly in good repute as mentioned by Evelyn. They are but little esteemed now, and it is doubtful if the varieties shown for them are true.

FOX KERNEL.—A middle sized, high coloured apple, ovate in shape, with angular sides. The tree bears its beautiful fruit very freely, and thus it has kept its place in the Herefordshire Orchards. It should, however, be sold in the market, for it has but a poor character as a cider fruit.

FOXLEY.—A seedling of Mr. Thos. Andrew Knight from the *Siberian Crab*, impregnated with the pollen of the *Golden Pippin*. It is a very small but beautiful apple, of a golden yellow colour, with a bright orange cheek. The specific gravity of the juice, Mr. Knight found to be 1·080. He thought it a very hardy and most valuable cider fruit, but it has failed to retain this character, and is but very little grown.

FRIAR.—A very old variety, formerly much esteemed. It is mentioned by Evelyn, and figured by Mr. Thos. Andrew Knight, who found the specific gravity of its juice to be 1·073. It has disappeared of late years, and was not exhibited in its true character.

GOLDEN BITTERSWEET.—A Devonshire apple, large and conical with ribbed sides. It is a yellow apple, with a red cheek, and sprinkled over with small russet dots and traces of russet. The tree bears well, and the fruit keeps well. It has a good repute as a cider apple.

GOLDEN MOYLE.—An apple grown on almost every farm around Ledbury. The tree grows large and bears well. The fruit makes good cider, and is also in high repute for the manufacture of jelly and jam. For this latter purpose the fruit, taken from the apple heaps, sold this year, (1884,) at four pounds the ton. "A sensible apple" the grower observed.

GOOSE APPLE.—A grass green apple, above middle size. It is very sour, cooks transparently, and makes excellent apple sauce—hence its name. The tree crops "wonderfully." It is chiefly used as a culinary fruit, but the remainder is welcomed at the cider press.

GRANVILLE.—A small red Somersetshire apple of good repute. It is supposed to give a high colour to the cider.

GREEN STYRE.—A middle-sized apple, late in season, and a good keeper. When it becomes yellow it is a good culinary apple, and is often sold as such. The tree is very large, and bears "tremendously." As a cider fruit it is also considered very good.

GRITTLETON RED.—A very good cider apple for a mixture of fruit, but has not sufficient character to be used alone. The tree is a great bearer.

GRITTLETON YELLOW.—Is a Gloucestershire apple of good repute in some districts.

GUINEA APPLE.—A small apple which looks like a crab, but is very sweet and luscious. It is chiefly found about Ullingswick, and the eastern side of the county. The fruit makes a rich red coloured cider of good character, and deserves to be grown more than it is.

HALL DOOR.—A large, red streaked apple, very conical in shape, with a projecting snout. The trees crop well, and the fruit sells readily in the market. This is its proper destination, for its qualities as a cider apple are but very moderate.

HANBURIES KERNEL.—A red-streaked apple, above middle size, good as cider or pot fruit. It was raised at Hanburies, in the parish of Bishop's Froome, and is spreading from thence in all directions.

HANGDOWN, or HORNER.—A small yellow apple, in high favour both in Devonshire and Somersetshire. The tree is small and spreading. It blossoms very late, not until June, and bears profusely. It is a late variety, and makes a good rich cider.

HARD-BEARER.—A second early apple, "something like *Skyrme's Kernel*, and quite as good." It is grown in the valley of the Froome river. The fruit has a bitter-sweet, astringent flavour, and makes excellent cider.

HATCHER.—A Gloucestershire apple, green and russety, with red streaks on the sunny side. The tree is middle-sized, and bears abundantly. It is a late variety.

HELLEN'S KERNEL.—A seedling raised by C. W. Radcliffe Cooke, Esq., M.P., at Hellens, Much Marcle (c. 1850). The density of the fresh juice is 1,057; it contains 12½ per cent. of sugar, but is very deficient in tannin, mucilage and salts. A good apple to mix with rougher varieties, but without sufficient character to make cider alone. It is a pretty fruit, and should be sold in the market for immediate use.

HEMING.—An old Gloucestershire apple mentioned by Evelyn, and formerly much esteemed. It is scarcely to be found now.

HOGSHEAD.—A very old variety mentioned by Forsyth. It is a small and astringent apple, but very juicy. It is considered very useful to mix with other and sweeter varieties.

HOLLOW-EYED PIPPIN.—An apple above middle size, very handsome, with angular sides. It is orange in colour with red streaks, and is most suitable for sale as table fruit. It maks a thin, poor cider.

HONEYCOMBE.—A Somersetshire variety. The tree is very vigorous in growth, and when full grown bears very abundantly. It makes a large, handsome tree; aud its fruit is said to make excellent cider.

IZARD'S KERNEL.—A variety somewhat similar to *Broad-tail*, but becoming more narrow towards the eye. It has also a much higher colour. It is grown about Ledbury, Pixley, and Aylton. It makes good cider, and is saleable as pot fruit when better varieties are scarce.

JERSEY CHISEL, CHISEL JERSEY, or BITTER JERSEY.—A striped bitter-sweet apple in the highest esteem in Somersetshire. It is a free grower and a constant bearer. It makes an excellent well flavoured cider, of high colour, and if mixed with some other sweet variety ripening at the same time, it becomes of the highest quality.

JERSEY FLENIER.—This is also a Somersetshire apple of good repute. The fruit is small, and red striped, with a juice of much richness and flavour. The tree bears profusely.

JONES' KERNEL.—A good looking apple, but its looks are deceptive. It is one of the very worst grown. "A single bushel would spoil a hogshead of good cider." The heads of the trees should all be cut off and regrafted with a better variety.

KILL-BOYS.—A green, middle sized Gloucestershire apple. The tree grows strongly with a drooping habit and bears freely. It is a late variety. Its acrid, rough tasted fruit has probably given it its name, as it also gives it its value for cider when mixed with other varieties of richer juice.

KNOTTED HEREFORD (formerly called *Knotted Norman*). A green, bittersweet apple, with a broad base, and more or less russety. The trees grow very knotty and knarled, and crop badly.

LANGWORTHY'S SOUR NATURAL.—A local Somersetshire apple of middle size. It is an early variety and bears well.

LANGWORTHY'S SWEET NATURAL.—A small red Somersetshire apple. It is also an early variety, but without much merit in any way.

MAGGIE.—A Gloucestershire cider apple of fair repute. It is a small, yellow apple, with a red cheek and sprinkled over with russet dots. The tree bears well, and the fruit has a very acid, austere taste.

MARROW-BONE or TOM PUTT.

MAUNDY, or PHILLIP'S MAUNDY.—A middle sized yellow apple, with a bright red cheek. It is second early. The fruit has a rough, astringent flavour, and is thought to give good keeping qualities to the cider from mixed fruits.

MONKTON.—A beautiful, small, red apple, raised at Monkton, near Taunton, in Somersetshire. It should be mixed with other fruits, since it has no decided vintage character of its own.

MORGAN'S SWEET.—A favourite cider apple in Somersetshire. It is a pale yellow, conical apple, with ribbed sides, and covered with dots. The tree grows well and bears freely. It is a late variety, and cooks well.

MORRIS' or MAURICE'S PIPPIN.—A Gloucestershire green russet apple of middle size. It is a late variety, and considered an excellent cider fruit.

MURDY APPLE.—A variety said to have been raised at Murdy, in Monmouthshire. It is a small bitter-sweet apple, rather soft, but very good and useful for cider. The trees are large and of upright growth, and bear well every second year The fruit is late, and its juice so rich that it will make excellent cider alone.

NATURAL POCKET APPLE.—A large Devonshire apple, much more useful as a culinary fruit than for cider-making. It is a handsome greenish yellow apple, with a red cheek and ribbed sides. It should always be sold in the market.

NETHERTON LATE BLOWER.—A Devonshire cider apple in much favour. It is a large, yellow, conical apple, with a pale red cheek and russety base. The tree bears freely, and the fruit keeps well. Its skin is so thick that birds will not injure the fruit.

NETHERTON NONSUCH.—A large, highly-coloured, and very handsome apple, presumably raised towards the end of the last century at Little Netherton, Dymock, Gloucestershire. There are here two very old trees, and many young fresh-grafted ones (1880). It is a heavy broad-based apple, with a deep eye. It is a good "all round" apple for dessert, culinary, or cider purposes. "It is a wonderful apple to run," and makes a pleasant but pale cider. It is a very useful, prolific variety, and the Messrs. Fawke, of Little Netherton, highly recommend it.

NEVER BLIGHT, LOPEN NEVER BLIGHT, or MORRIS' APPLE.— A round middle-sized apple of high colour. The tree is very hardy, and a great bearer, scarcely ever failing to produce a crop. It has a sweet rich juice, and is considered an excellent cider apple.

NEW BROMLEY. — A small bright-coloured apple, much esteemed in Gloucestershire as a cider fruit. Its flesh is often tinged red, and its juice has the astringency so useful with cider fruits.

NORTHWOOD BITTERSWEET.—A large Somersetshire apple, white and red striped. The tree is large and generally bears well. It is sold chiefly as a table fruit.

OAKEN PIN.—An old variety mentioned by Evelyn. The fruit of this name in Devonshire is large, and sells well as a cooking apple. This, however, is not a rich cider apple, and is not the old variety known by this name.

OATLAS KERNEL, or OATLEY'S KERNEL.—An apple of middle size, and of a pale green colour, streaked with red. It is an old variety grown at the Frith Farm, in the parish of Ledbury, and in some of the surrounding orchards. It is considered a good cider apple, and is useful for table purposes when required.

OLD GERMAIN, or OLD JARMAN.—A large good looking apple which keeps and cooks well. Its proper place is the market, and not the cider mill.

OLIVE.—A variety mentioned by Evelyn, and said to grow near Ludlow. It has not kept favour in modern times.

ORANGE PIPPIN.—A very beautiful apple, like the *Blenheim Orange*, but smaller, and more regular in shape. It makes good cider, but usually finds its way to the market, where its beauty commands for it a ready sale. The tree grows well and blossoms well, but it is a shy bearer, and a good crop can only be looked for once in every four or five years.

OTLEY.—A Shropshire apple formerly held in great esteem. Phillips' says of it :—

> " Salopian acres flourish with a growth
> Peculiar, styl'd the Otley : Be thou first
> This apple to transplant : if to the Name
> Its Merit answers ; nowhere shalt thou find
> A wine more priz'd, or laudable of Taste,"

The poet's advice, however, does not seem to have been followed.

PAWSAN.—An old variety, mentioned by Phillips, and figured by Mr. T. A. Knight in the "Pomona Herefordensis." He found the specific gravity of its juice to be 1.076. The name appears at our shows, but not the true apple.

PIN APPLE.—A local apple of good repute. The original tree at Much Cowarne has an iron pin driven through it, to prevent a split from spreading—hence its name. It is a round, green and yellow apple, late in season, and makes a very good cider without other varieties.

POOR MAN'S PROFIT.—A small, striped Somersetshire apple, a late variety, which is thought to make very good cider.

POUGHILL GREEN.—A large green Somersetshire apple, which keeps well. It only finds its way to the cider mill when the crop is abundant, and the market overstocked.

POUND APPLE.—A very large apple without sufficiently good qualities to keep it in the market, and it is used therefore in Devonshire and Somersetshire for cider. It quickly fills the cask, but requires apples of better character to give strength and flavour to the liquor.

PREECE'S KERNEL.—A large apple, which ripens early and decays quickly. It has little merit, either on the table, or in the cider press.

PRICE'S BITTERSWEET.—A late apple, striped red and green, rather below middle size. It is thought one of the best apples in the Froome valley, and makes excellent cider alone, or in mixture.

PUPPY SNOUT.—A middle sized apple of narrow pointed shape. It is late in season, and of rather doubtful character as a cider fruit.

RAMPING TAURUS.—A recent variety, grown at Fair Oaks Farm, Castle Morton, Worcestershire. The fruit is large, conical and angular, greenish white, and bittersweet. It makes "grand cider" and very strong. This apple has the peculiarity of baking well, but it will not boil.

RED CLUSTER.—A small red Somersetshire apple, a late variety, which gives excellent assistance in making cider from mixed fruit. The tree bears freely.

RED MUST, or MUSK.—This is the largest cider apple grown in Herefordshire, and is therefore seldom used as such. It has a light thin juice, of the specific gravity 1.064 (Knight), and is not so much esteemed now as it was formerly.

RED SOLDIER.—"A very lucky bearer," and from this, and its bright colour, it was much sought after a few years since. However, it only makes a thin, poor cider, and has thus lost its repute. It should be sold in the market, where a good colour sells anything.

RED STYRE.—A small apple, almost entirely covered with dark crimson. It is an excellent cider fruit, and highly valued in the Froome valley, where it is chiefly to be found.

RED TURK, or BLOODY TURK.—An early, soft, deep red apple, the colour extending more or less through the flesh. It is a bad keeper and a poor cider fruit. It, too, should be sold to the costermonger.

RED WILDING.—A late apple of middle size. Its juice does not fine well, and it is only useful to mix with other varieties.

REYNOLD'S CRAB, or RAYNAL'S CRAB.—A yellow fleshed fruit, with something of the flavour of the *Siberian Crab*. The tree grows to a large size, and bears "wonderfully." The fruit makes "the very best cider."

RUSTY COAT.—A Gloucestershire apple of good repute. It is a small yellow apple, with an orange cheek, specked and marked with rough russet. It is a late fruit, and thought to make excellent cider.

SEA SPAWN.—A local variety from Dilwyn, very small in size. The tree bears very freely, and the fruit is thought to add virtue to mixed fruits.

SHEEPS SNOUT, or SHEEPS NOSE.—A light, green, bitter sweet apple, largely grown in Somersetshire, Gloucestershire, and Worcestershire. It is of medium size, and of a somewhat narrow, oblong shape, with sharp angles. It is valued as a cider fruit, and cooks well when in season.

SIBERIAN BITTER SWEET.—A very handsome, small, globular fruit, of golden colour, with a red cheek, growing in clusters. It is a seedling of Mr. Thomas Andrew Knight's, produced from a seed of the *Yellow Siberian Crab* fertilized with the pollen of the *Golden Harvey*. The juice is sweet, without acidity, with the high specific gravity of 1.091. It has failed, however, as a cider apple, but is very useful for making preserve, or jelly.

SIBERIAN HARVEY.—Another seedling of Mr. Knight's, from the same parentage as the last named apple, and its juice has the same high specific gravity 1·091. It first fruited in 1807, when it obtained the annual premium of the Herefordshire Agricultural Society. It is a beautiful fruit, growing in thick clusters. Mr. Knight thought it would prove to be a cider apple of the highest merit, but it has not gained this character, and is now but little grown.

SLACK-MY-GIRDLE, or SLACK-MA-GIRL.—A striped Somersetshire apple of large size. It keeps well, and is usually sold for culinary purposes, though it often helps to fill the barrel. As a cider apple, however, it has not much merit.

SOPS IN WINE.—An apple above middle size, orange red on the shady side, and very dark red towards the sun. The fruit has a bloom on the surface. The flesh is also coloured red, more or less. The tree is large and bears well. It is considered a good culinary and cider fruit.

STEAD'S KERNEL.—An ovate, conical apple of middle size. It was raised by Mr. Daniel Stead, of Brierley, near Leominster. It is a late variety, yellow in colour, with specks and lines of grey russet. It is a valuable bitter sweet cider apple, with a combined sweetness and astringency. Its juice has the specific gravity of 1·074 (Knight). It cooks well during its season.

STYRE, or SMALL STYRE.—A small red apple of oblong shape, and yellow flesh. It makes excellent cider. The apples look like plums on the tree.

SUGAR APPLE, or SUGAR LOAF.—A pot, or cider fruit, grown on every farm in the parish of Ledbury and its neighbourhood. It sells well in the market, but it "helps to make first class cider, and for this it is always kept by those who know its virtue."

SUGWAS KERNEL.—A local variety grown at Sugwas, near Hereford, but without any very great merit.

SUSSEX APPLE.—A Sussex pippin, hard in texture, and covered with brown russet. It has a rough, harsh taste, and is a good cider apple. The tree is not "lucky" in bearing.

SWEET RENNET, or REINETTE.—A green Somersetshire apple, of middle size. It is an early variety, and bears well, but has not sufficient character to make good cider by itself.

TANKERTON.—An apple of full middle size, white, with a pink cheek. The tree grows thick in the wood, and bears well. It is a mid-season apple, cooks well, and makes a fair cider.

TEN COMMANDMENTS.—A deep red, rather conical apple, with ribs, becoming very prominent near the eye. The flesh is white, stained here and there with red. When cut across, it shows ten red spots around the core, and hence gets its name. The tree bears well, and the fruit is thought to make good cider.

TRACE APPLE, or TRACED HEREFORD (formerly called *Norman*).—A Herefordshire seedling, which bears freely, and keeps well, but which is without any very special merit as a cider apple.

TREMLETT'S BITTER.—A Devonshire bittersweet apple, above middle size, and highly esteemed as a cider apple.

TURK'S CAP.—A large orange yellow apple, sprinkled with grey dots. It has an acid, astringent taste. It is usually sold for culinary purposes, but often finds its way to the cider mill.

UNDERLEAF (HEREFORDSHIRE).—A green middle sized apple that may serve for table or cider fruit. The tree is large, the wood grows thickly, and the leaves conceal the fruit, and thus it gets its name. It is a good keeping apple, and usually finds its way to the market, but is nevertheless considered also a very good cider apple.

WELL BELOVED.—A large handsome second early apple, which sells well in the market as pot fruit. It bakes well, but as a cider fruit it has not much merit.

WHITE GRAPES, or WHITE CLUSTER.—A small, white Somersetshire apple. The tree bears profusely and is therefore a good cask filler, which is its chief merit.

WHITE MUST, or MUSK. A small fruit of a pale straw colour. The gathered fruit quickly becomes unctuous to the feel and has a peculiar ether like smell. Its flesh is so soft that the least touch bruises it. It makes a thin, pleasant, cooling drink for the hop pickers. It will also cook well.

WINTER POOL.—A large oblong apple, which may be used for either table, or cider fruit, but is not of high quality in either case. The tree moreover is a bad bearer.

WITHINGTON RED, or REDSTREAK.—A pretty apple, rather below the middle size. The tree bears well, but the fruit has no very special merit as a cider apple.

WOODCOCK.—A very old variety mentioned by Phillips, and figured by Mr. Knight in the "Pomona Herefordensis." It was formerly held in great esteem and its juice had the specific gravity of 1·073, but it has disappeared from our orchards of late years, and the fruit exhibited at the apple shows has not been true to character.

WOODSELL.—An old variety of high repute. It is still grown at Much Marcle, and here and there in the South Eastern side of the county. Its cider, in a fine season, is said to be "as good as Foxwhelp." It is certainly a valuable variety, and one that merits more extensive cultivation.

YELLOW STYRE.—This is a very excellent cider fruit. It is grown more in West Worcestershire, at Bushley, Chaseley, Upton, &c., than in Herefordshire. The trees that yet remain are very old, and young ones have not been grafted. It well deserves further propagation.

GENERAL INDEX.

	PAGE
Acetic Fermentation	53
Agricultural Return of British Orcharding	73
Alcohol in Cider and Perry ...	64
Alford, or Sweet Alford Apple ...	225
American Blight	34
Amphlett's Favourite Apple ..	226
Ansell, or Ancell Apple	226
ANTIFERMENTS	68
Analysis of Apple Juice	48
————— Apple and Pear Trees	16
————— Credenhill Soil	11
————— Fruit Sugar	49
————— Norman Apples ...	90
APPLE CONGRESS AT ROUEN ...	28
Apple Heaps	40
Apple Mill	42
Apple Trees, Life of	25
Apples, Norman, introduced 1884	89
Apples, Old Varieties	23
————— New Varieties	26
ARGILE GRISE Apple	91
ARLINGHAM SQUASH Pear ...	177
AYLTON RED Pear	179
BARLAND, or Bareland PEAR ...	180
BASTARD FOXWHELP Apple ...	92
Bastard Rough Coat Apple ...	226
Baylis' Kernel Apple	226
Beale, Dr.	6
BÉDAN-DES-PARTS Apple ...	93
Bell, or Bell Norman Apple ...	129
Belle Orchard Seedling Apple ...	226
Bennett Apple	226
Best Bache, or Bache's Kernel Apple	226
Bitter Jersey Apple ...	233
Bitter Scale Apple ...	226
Black Bud Apple	226

	PAGE
Black-eyed Pippin	227
BLACK FOXWHELP Apple ...	94
BLACK HEREFORD Apple ...	96
Black Horse Pear	224
BLACK HUFFCAP Pear	183
Black Kingston Apple	132
Black Norman Apple	227
Black Wilding Apple	227
BLAKENEY RED Pear	184
Bloody Turk Apple	238
Bosbury Pear	180
Bosbury Scarlet Pear	223
Bottle Stopper Apple ...	227
BRAMTOT Apple	97
BRAN ROSE Apple	98
Bridge Pippin	227
Bristol Crab	227
Broad-eyed Pippin	227
Broad-leaved Hereford Apple ...	227
Broad Tail Apple	227
Bromesberrow Crab	228
BROMLEY Apple	99
Brownseys Apple	228
Brown Huffcap Pear ...	183
Brown Snout Apple	228
BUDDING AND GRAFTING ...	23
Bull's Eye Apple	228
BUTT PEAR	186
Cabbage Apple	228
Cadbury Apple	154
Candid Heart Apple	228
CANKER OF APPLE TREES ...	33
Canon Apple	228
Canon Bittersweet Apple ...	228
Captain Nurse Apple, or Captain's Kernel	228
CARRION Apple	101
CATALOGUE OF (ROUEN) APPLES	90

INDEX—continued.

	PAGE		PAGE
Chaxhill Red Apple	228	Dymock Red Apple	111
Chaseley Green Pear	187	———— Pear	224
Cheatboy Pear	189	Eggleton Styre Apple	113
Chemical Analysis of		English Cider and Perry Orchards	2
Apple and Pear Juice	48	Essex Kernel Apple	230
———— Apple and Pear Trees	16	Estimation of Sugar by Density	49
———— Credenhill Soil	11	Excels Apple	230
———— Fruit Sugar	49	Farmer Hearland's Apple	230
Cherry Hereford Apple	102	Fawkes' Kernel Apple	230
Cherry Pearmain Apple	103	Fermentation, Acetic	53
Chibble's Wilding Apple	229	———— Active, Dilatory, or Persistent	66
Chisel Jersey Apple	233	———— Cleanliness in	46
Cider Apples, Old Varieties	23	———— Insensible	71
———— Modern Varieties	26	———— Pasteur's Theory of	51
Cider Brandy Apple	229	———— Practice of	55
Cider Making	58	———— Process of	57
Cider and Perry Orchards	2	———— Process of American	61
———— Commercial Value of	74	———— Process of French	61
———— in Bottle or Cask	77	———— Process of Jersey & Channel Islands	61
———— Preservation of	70	———— Putrid	54
Cider Merchants	4	———— Viscous	53
Cider Lady's Finger Apple	104	Fillets, or Violets Apples	231
Cider preferred by Kings	6	Fining Cider or Perry	70
Claret Wine Apple	229	Forest Pear	223
Coccagee Apple	106	Forest Styre Apple	114
Coleing Apple	229	Fox Kernel Apple	231
Commercial Aspect of Orchards	73	Foxley Apple	231
Cook's Kernel	229	Foxwhelp Apple	116
Coppy Pear	190	Foxwhelp recultivated	26
Corn Apple	229	Fréquin Audiévre Apple	121
Cowarne Red Apple	108	Friar Apple	231
Cultivation of Old Apples	26	Fruit Management	37
Cummy Apple	109	———— Gathering	38
Darbin Red Streak Apple	229	———— Grinding	44
De Bouteville Apple	110	———— Heaping	40
Dean's Apple	229	———— In Mill	42
Deterioration of Orchards	4	Fruit Tree Acreage	73
Devonshire Red Streak Apple	229	Fruit Tree Enemies	32
———— Royal Wilding	229	Fungus Growths	35
Difficulties of Fermentation	66	Fungus Yeast Plants	51
District Cider Factories	82	Garter Apple	122
Dufflin Apple	230	Gennet Moyle Apple	124
Dunn's Beloved Apple	230		
Duration of Apple Varieties	25		

INDEX—*continued*.

	PAGE		PAGE
Gin Pear	224	Knock Down Pear	224
Golden Bittersweet Apple ...	231	Knotted Hereford Apple ..	234
Golden Moyle Apple	231	KNOTTED KERNEL	134
Goose Apple	231	La Belle Normande	129
GRAFTING AND BUDDING ...	23	Langworthy's Sour Natural Apple	234
Granville Apple	232	———— Sweet Natural	
Greasy Apple	137	Apple	234
Green Squash Pear of Evelyn...	177	Le Cidre	5
Green Styre Apple	232	Lichens and Mosses	36
GREEN WILDING Apple	126	Local Perry Pears	223
Grittleton Red Apple	232	LONGLAND, or Longden PEAR ...	193
———— Yellow Apple ...	232	Long Stalk Pear	224
Guinea Apple	232	Lopen Never Blight	235
HAGLOE CRAB	127	Lord Scudamore's Crab	146
Hall Door Apple	232	Lumberskull Pear	224
Hanburies Kernel...	232	Maggie Apple	234
HANDSOME HEREFORD (or Norman)	129	Malic Acid...	50
		Malvern, or Malvern Hill Pear	194
Hangdown, or Horner Apple ...	232	Marrow Bone Apple	234
Hardbearer Apple	232	Maundy Apple	234
Hartpury Green Pear	187	Maurice's Pippin	234
Harvest Apple	229	MEDAILLE D'OR Apple	135
Hatcher Apple	232	MICHELIN Apple	136
Hellen's Kernel	233	Mildew	36
Heming Apple	233	Mistletoe	32
Herefordshire Norman Apples...	29	Monkton Apple	234
HISTORY OF CIDER AND PERRY ORCHARDS	2	Morgan's Sweet Apple	234
		Morris Apple	234
Hitterly Apple	102	MOORCROFT PEAR	194
Hogshead Apple	233	Mosses and Lichens	36
Hollow-eyed Pippin	233	Mucilage in Juice	50
HOLMER, or Holmore PEAR ...	191	MUNN'S RED Apple	137
Home Fruit Markets	85	Murdy Apple	235
Honeycombe Apple	233	Natural Pocket Apple	235
Hybridization of Fruits... ...	21	Netherton Late Blower Apple...	235
Insect Blights	34	———— Nonsuch Apple ...	235
Irchinfield Red Streak Apple ..	146	NEWBRIDGE PEAR	196
Izard's Kernel	233	New Bromley Apple	235
Jersey Chisel Apple	233	Never Blight Apple	235
Jersey Flenier Apple	233	NEW MEADOW PEAR	198
JOEBY, or Joby CRAB	130	NORMAN CIDER APPLES, introduced	88
Jones' Kernel Apple	233		
Kempley Red Apple	101	Northwood Bitter Sweet Apple	235
Killboy's Apple	233	Nurse's Kernel	228
KINGSTON BLACK Apple ...	132	Oaken Pin Apple...	235
Knight, Mr. Thomas Andrew ...	20	Oatlas, or Oatley Apple	236

246   INDEX—*continued.*

	PAGE		PAGE
Old Bromley Apple	99	Prizes at Rouen Congress	87
Old Germain, or Jarman Apple	236	PROCESS of FERMENTATION	57
Old Red Sandstone Soil...	11	Puppy Snout Apple	237
Old varieties of Orchard Vintage fruits	23	Putrid Fermentation	54
		PYM SQUARE Apple	138
OLDFIELD PEAR	199	Ramping Taurus Apple...	237
Olive Apple	236	RED BUD Apple ...	140
Orange Pippin	236	Red Cluster Apple	237
ORCHARD AND ITS PRODUCTS	1	RED FOXWHELP APPLE...	141
Orchard—Aspect and Site	14	Red Must, or Musk Apple	237
——— Authorities	8	RED, or Red Horse, PEAR	206
——— Brandy ...	65	RED HEREFORD Apple ...	142
——— Budding and Grafting	23	Red Norman Apple	142
——— Commercial Aspect of	73	RED ROYAL Apple	144
——— Culture ...	6	Red Soldier Apple	237
——— Encouragement of	84	Red Spider...	35
——— Manure...	15	RED SPLASH Apple	145
——— Planting...	19	Red Squash Pear .	210
——— Pruning ...	30	REDSTREAK Apple	146
——— Prospects	83	Red Strake of King's Caple	146
——— Renovation of ...	78	Red Styre Apple ...	237
——— Seedlings	20	Red Turk Apple ...	238
——— Soil	9	Red Wilding Apple	238
——— Surface and Drainage ..	12.	REJUVENATED FOXWHELP Apple	150
——— Trees, and Varieties of	20	RENOVATION OF ORCHARDS	78
Otley Apple	236	Report on Rouen Congress	87
PARSONAGE PEAR ..	201	Reynold's or Raynal's Crab	238
PASTEUR'S THEORY OF FERMENTATION	51	ROCK PEAR	207
		Rouen Catalogue of Fruit for the Press	90
Pawsan Apple	236	ROUGE BRUYÈRE Apple ...	153
Perry, First made	2	Royal Cider	65
Perry, Manufacture of ...	62	ROYAL WILDING Apple...	154
Phillips' Maundy Apple...	234	Rust in Orchards ...	36
Pin Apple ...	236	Rusty Coat Apple	238
Pine Pear ..	202	SACK APPLE	156
Pint Pear ...	204	Sack Pear ...	179
Pomage	56	Saccharometer	79
POMOLOGICAL CONGRESS AT ROUEN ...	28	Salicylic Acid	69
		SAM'S CRAB APPLE	157
Poor Man's Profit Apple	236	Scudamore's Crab ..	146
Poughill Green Apple	237	Sea Spawn ...	238
Pound Apple	237	Seedling Apples ...	20
Pot Fruit ...	81	——— Pears	22
Practice of Fermentation	55	SKYRME'S KERNEL	159
Preece's Kernel ...	237	Sheep's Snout Apple	238
Price's Bittersweet Apple	237		

	PAGE		PAGE
Siberian Bittersweet Apple	238	Test for good Apple Trees	79
—— Harvey Apple	238	THORN PEAR	212
Slack-my-girdle or Slack-ma-girl	239	THURSTON RED Pear	214
Soil	9	Tump Pear	223
Sops in Wine Apple	239	Trace, or Traced Norman Apple	240
Sow Pear	224	Tremlett's Bitter Apple	240
Small Styre Apple	239	Turk's Cap Apple	240
SOUTH QUEENING Apple	161	Turner's Barn Pear	224
Specific Gravity of Juice	48	Underleaf (Herefordshire) Apple	240
SPREADING REDSTREAK Apple	163	UPRIGHT RED STREAK Apple	168
Staunton, or Stanton Squash Pear	217	Value of Herefordshire Orchards	74
		Vegetable Blights	36
Stead's Kernel Apple	239	Violets or Fillets Apples	231
STONY-WAY PEAR	209	Viscous Fermentation	53
STYRE, Stire, or Styrom	114	Well-beloved Apple	240
—— Eggleton Apple	113	White Cluster Apple	240
—— Forest Apple	114	White Grapes Apple	240
—— White Apple	172	WHITE HEREFORD Apple	169
—— WILDING Apple	165	White Horse Pear	216
STRAWBERRY HEREFORD (late Norman)	164	White Longland Pear	216
		WHITE MUST, or MUSK Apple	170 / 240
Sugar, or Sugar Loaf Apple	239		
Sugwas Kernel Apple	239	White Norman Apple	169
SULPHUR, as an antiferment	68	WHITE SQUASH Pear	217
Sussex Apple	239	WHITE STYRE APPLE	172
Sweet Rennet, or Reinette Apple	239	WILDING BITTERSWEET Apple	173
Table Fruit	81	WINNALL'S LONGLAND Pear	219
Tankerton Apple	239	Winter Pool Apple	240
TANNER'S RED Apple	167	Withington Red Apple	240
TANNIN	49	Woodcock Apple	240
Tartaric Acid	50	Woodsell Apple	241
TAYNTON, or Tainton, SQUASH Pear	210	YELLOW HUFFCAP Pear	221
		YELLOW REDSTREAK Apple	174
Tainton, or Taynton, Black Apple	132	Yellow Styre Apple	241
Ten Commandments Apple	239	YOKEING HOUSE Pear	222

www.ingramcontent.com/pod-product-compliance
Lightning Source LLC
Chambersburg PA
CBHW021353230426
43666CB00006B/507